NEW THEORY
OF
TRISECTION

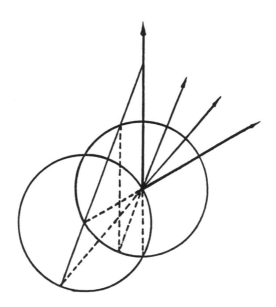

NEW THEORY
OF
TRISECTION

SOLVED THE MOST DIFFICULT MATH PROBLEM
FOR CENTURIES IN THE HISTORY OF MATHEMATICS

FEN CHEN

INTERNATIONAL
SCHOOL MATH & SCIENCES
INSTITUTE
ALEXANDRIA, VIRGINIA, U.S.A.

Fen   Chen
International
School Math & Sciences
Institute
P.O.Box 16707
4520 King Street, #902
Alexandria, Virginia 22302
U.S.A.

AMS Classifications: A1-A05

Library of Congress Cataloging-in-Publication Data

Chen,   Fen
  NEW THEORY OF TRISECTION
  1. 146 Drawings in the New Thinking of Trisection. 1. Title
  VAU 337-290   1995   Size: $6\frac{1}{2}$' x $9\frac{1}{4}$'

  Library of Congress Catalog Card Number is: 99-094277

  ISBN:0-9671511-0-4

Printed and bound by GOODWAY GRAPHICS, Springfield, Virginia, U.S.A.
Printed in the United States of America (U.S.A.)

Preface

The famous trisection problem has been caught in my mind since I started learning mathematics in Junior High School. My geometry teacher challenged me to solve the problem in a classroom. But, at that time, he did not xplain clear the nature of trisection. However, I keep trying this problem over in my mind year after year.

One day after I had graduated from Taiwan Normal University became a mathematics teacher in a public school. An expert architect of a traditional temple builder came to visit me for examining his trisection work.. He had showed his talent without formal educational training. His trisection was drawn and constructed by mean of curve which had been discovered in early time. But, the trisection problem had refreshed my mind again. At that time, I did not devote and concentrate to solve or rethink the problem because I was very busy in my teaching.

After I had studied mathematics education in the graduate school at Tokyo University of Education in Japan, my thinking mind an philosophy of education had deeply impacked by a famous Japan mathematics educator Professor Yushi-Nobu Wada. He always emphasized five characteristics of mathematics education that are: solving, reasoning, thinking, discovering, and appling abilities. If a problem is soluble, the solution of the problem must be examined and verified as to the accuracy. If the problem is unsoluble, we must find the main answer for the question:"Why is the problem unsoluble?" The basic fundamental reason needs to be explained clearly without any ambiguity.

Before Christmas of the year 1989, I went to teach mathematics at Walter Whiteman High School in Montgomery County,Maryland. It was an honor geometry class. The subject for the class was geometric construction. After we had discussed the problrms in the content, the students challenged me to explain the famous

trisection problem and I started writing this theory. My
students asked me the question:"Why is the trisection of an
angle impossible?" They took several particular angles such
as; π angle, $\pi/2$ angle, and $\pi/4$ angle can be trisected into
three equal angles. Why can't we trisect an arbitrary angle?
At that moment, I responsed to the students: " I will rethink a-
bout the problem and find the main reason." At the time, a
new idea came into my mind, we can draw a Christmas tree by
constructing two versions of the angle, then, three versions
of the angle, four versions of the angle, --- and so on. I
drew the figure on the blackborad (see the construction on
the appendix I.) The entire class was very interested in the
construction.

The Christmas tree started me going over the idea keenly.
We can start one given acute angle to construct three times
angle. Reversely, from the three times angle, we go back to
find the original given acute angle getting along with two
equal circles. The idea had been developed to invent a new
trisector (see the United Stated Patent Office awarded me the
patent No. 5,210,951 on the appendix II.) After I received on
the trisector recognition had brought me more confidence to
focus on the new theory of trisection.

Actually, the first drafting was completed in the year 1989.
Then, I revised my manuscript in 1991. For the years, I was
continuing to draw 146 constructions which the United States
Copyright Office rewarded me the copyright No. VAu337-290 on
March 27, 1995. Those constructions are used to test the eleven
theorems in the entire discussion, each trisection is following
an unique pattern. Though the years, I have input and devoted
myself to mathematics education, the trisection problem is one
of my works. I want to publish and share my work with all the
reluctants of mathematicians, educators, scholars, engineers,

achitects, and amateurs. I believe that the entire content
of this work has based and developed faithfully on the basic
logical reasons and the original geometric manner and method
of the Euclidean Geometry as a mathematical model.

I would like to take this opportunity to say:"Thank you very
much indeed with my heart." First, Professor T.A. Romberg at
National Center for Research in Mathematics Science Education
of the University of Wisconsin-Madison. He has make the most
positive encouragement for me to keep going on searching the
key points of trisection. It was an important turning point
for correcting my original gesture of the Theorem III in the
Chapter III. Second, Professor Robert Mozon at the University
of Kentucky, he was working at Nation Science Foundation, when
I mailed him my first manuscript, he responsed very sincerely
comments :" When mathematicians say that it is impossible to
trisect an angle in an algebraic manner." on the April 18,1991.
Clearly, professor Molzon had make hint that trisection could
be done by the geometric manner. Therefore, I kept reexaming
and testing the eleven theorems in the entire theory. Third,
Mr. Hampton Williams at James Madison Hight School, Fairfax
County Publics Schools, Virginia. He is an excellent senior
teacher in instructing mechanical drawings for thirty years
experience. He helped me test 50 degrees, 60 degrees, and 90
degrees skillfully, as the result, each angle was trisected
precisely and exactly in terms of the construction method of
theory. Those testings are showing on the pages 12, 13, and
14. Fourth, my friend Professor Hunter Alexander helped me
read and cite my manuscript several times. His contribution
was remarkable.

Finally, I would like to thank the many people who have
helped and instructed me at various time from Japan to the
United States of America. In the University of Maryland -
College Park, they are: Professor Jim Fry, Professor James

Preface

Henkelman, Professor Niel Davidson, Professor Martin Johnson
Professor Partricia F. Campbell, Dr. Elizabeth Shearn, and
Pastor Brian D. McLaren.

In the University of Wisconsin-Madison, they are: Professor
T.A. Romberg, Professor Elizabeth Fennema, and Professor Tom
Carpenter. Also, I would like to thank you very much to my
friends Pastor Robert Paulson, Ms. Margret Baungartner, and
Mrs. Mago Redmond, Mr. Denial Hardy.

In the University of Michigan, English Language Institute-
Ann Arbor, they are: President Dr. Harold Shapiro, Director
Dr. Larry Selinker, Secretary Dr. R.L. Kennedy, Dr. Eleanor
Foster, and all instructors.

In the Tokyo University of Education-Tokyo, Japan, they are:
all the members of Kyu-Dai-Ken Mathematics Education Study Group.
The study group was found by my Professor Yushi-Nobu Wada. It
is a leading group in Japan. All the outstanding mathematics
educators are joining the study group in Japan. They always
have research meetings weekly, monthly, and yearly in-service
mathematics teacher education. Professor Wada past away in
three years ago, but his research spirit is always keeping alive.
Especially, this work NEW THEORY OF TRISECTION is to memorize
Professor Wada and appreciate all the members of Kyu-Dai-Ken.

I am also very grateful of my brothers, sister, and nephews
in my native country. Mr. Bill Badmer's family and Mr. Tom
Shumaker's family in the United States of America. None of this
would have been possible to publish this research.

Thank you very much again.

                                        Fen   Chen
                                        Alexandria, Virginia
                                        U.S.A.

CONTENTS

SYMBOLS

$\angle$	angle
$\lvert x \rvert$	absolute value of x
$\odot$	circle
$\overset{\frown}{AB}$	the arc of two points A and B on a circle
$=$	equal or equal to
$\neq$	is not equal to
$>$	is greater than
$\geq$	is greater then or equal to
$<$	is less than
$\leq$	is less than or equal to
$m\angle ABC$	is measurement of an angle ABC
$m\,\overset{\frown}{AB}$	is measurement of an arc $\overset{\frown}{AB}$
$\overleftrightarrow{AB}$	the line through two points A and B
$\overline{AB}$	the segment with two end points A and B
AB	the length of the segment $\overline{AB}$
$\parallel$	is parallel to
$\square$	parallelogram
$\perp$	is perpendicular
$a : b$	the ratio of a and b
$\triangle$	triangle
$\square$	rectangle
$\overrightarrow{OA}$	ray OA the union of segment $\overline{OA}$ and the set of all points such that A lies between each of the points and O. O is called the end or initial point of the ray

INTRODUCTION

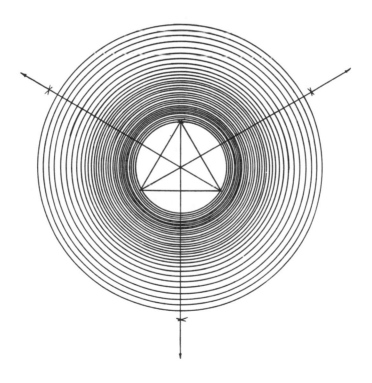

Questions:

1. Why a right angle can be trisected perfectly and logically
    by using an unmarked straight-edge and compass ?

2. Can we generalize the trisection of a right angle to trisect
    an arbitrary acute or obtuse angle ?

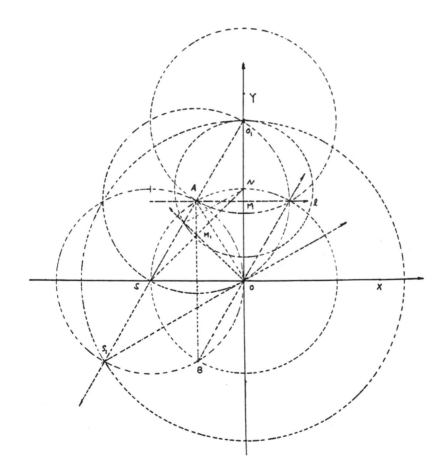

Knowledge of the developing mathematics or science of technology
is to discover new theories or technical skills for solving a very
difficult problem. To solve a vigorous problem is one major source
of mankind's progression. In the past history of mathematics, three
intricate problems were proposed by the ancient Greeks. These rigid
problems are: Trisection of an angle, duplication of the cube, and
squaring the circle. The three propositions are now called famous
problems because of their hardy solution. Especially, the trisection
problem was challenged by scholars, mathematicians, engineers, and
amateurs for over twenty-five hundred years. The constructing tools
are limited to an unmarked straight-edge and compass.

In the year 1837, a young German mathematician Pierre Laurent
Wantzel (1814-1849) proved there is no ruler-and -compass to divide
an arbitrary angle into three parts. The proof was stated on the
Wantzel's work: "Recharches Sur Les Mogens De Reconnaîrtre Si un
problème De Géométrie." Wantzel's proof of his theorem is based
upon the following two principles:

1. A given angle α can be constructed if and only if it is possible
   to construct a line segment or length $\cos\alpha$; and therefore from
   what we have said in squaring a circle. It follows that an
   angle α can be constructed if and only if the number $\cos\alpha$ can
   be constructed using the four operations of arithmetics; +, -,
   ., ÷ and the square root ($\sqrt{}$) operation.

2. From the well-known trigonometric identity
   $$\cos 3\beta = 4\cos^3\beta - 3\cos\beta .$$
   It follows that for any angle α we have let $\alpha/3 = \beta$. Thus,
   $\cos\alpha = 4\cos^3(\alpha/3) - 3\cos(\alpha/3)$. Now, let be any given angle
   and define a = $\cos\alpha$, and x = $\cos(\alpha/3)$. We can trisect the
   angle if only if the number $\cos(\alpha/3)$ is constructible, and
   therefore we can trisect the angle , if the roots of the
   algebraic equation: $4x^3 - 3x = a$ are constructible numbers.

Why rethink and investigate the trisection problem?

There are at least three reasons. Why the trisection problem was studied by so many mathematicians, engineers, architects, scholars, and amuteurs for over twenty-five hundred years. In the year 1989, Underwood Dudley wrote: <u>A BUDGET OF TRISECTIONS</u>. He had collected and illustrated 132 remarkable works on the trisection problem.

The first reason is that Wantzel's proof was the transfer of the problem from the realm of pure geometry into that of Algebra and Arithmetic. Also, wantzel used the operations of Arithmetic such as " + ", " - ", " · ", " ÷ " and square root ($\sqrt{}$) operation beyound non-quantitative point sets. The essential and originial Euclid's Elments are to be constructed by points and lines. The knowledge of Euclid's Elements is not only to measure the size of a segment or an angle or an area or an arc, but also to find the properties among figures. As we know, a segment can be measured by its length, but a ray or a straight line canot be measured its length. Therefore, the first principle of Wantzel's proof is not based on the nature of Euclid's Geometry in a mathematical model.

The second reason is the definition and its measurement of an angle. An angle is defined by two sides and one vertex on the same geometric plane. Theoretically, the measurement of an angle is the length of arc with center at the vertex of circle, also, we define the unit of arc is Pi (π) radian. Here, the magnitude of an angle is not a rational number. It is a transcendental number which is not a normal terminating or repeating decimal. Hence, if we are going to trisect an angle we must define a point functions to form the identity under single and triple angles. As we know, we defined six trigonometric functions as the ratio two sides of a right triangle. The unit of ratio among the six functions is not a radian. The value of the ratio is indicated by a number without a point on an arc.

The third reason is the solution of the equation: $4x^3 - 3x = a$ of the secondary principles. In the sixteenth century, an Italian physician and mathematician Jerme Cardan (Girolamo Cardano) (1501-1576) had solved a cubic equation: $x^3 + px + q = 0$, a solution of the reduced cubic equation by the substitution $x = u + v$ which will be a root of the cubic equation: $x^3 + px + q = 0$, if $u^3 + v^3 = -q$, and $uv = (-1/3)p$ or if u is a root of the following quadratic equation: $(u^3)^2 + q(u^3) - p^3/27 = 0$ and $uv = (-1/3)p$. If u^3 is a cube root of $\frac{1}{2}(-q + \sqrt{q^2 + 4p^3/27})$, and $v_1 = (-1/3)(p/u_1)$, then the three roots of the reduced cubic are:

$$z = u_1 + v_1, \qquad z = wu_1 + w^2v_1, \qquad z = w^2u_1 + wv_1$$

Here, $w = -1/2 + 1/2.\sqrt{3}i$ ($i = \sqrt{-1}$) is a cube root of unity. This is equivalent to the formula:

$$x = (-1/2.q + \sqrt{R})^{1/3} + (-1/2.q - \sqrt{R})^{1/3}$$

And, $R = (1/2.q)^2 + (1/3.p)^3$

According to the Cardan's solution a cube equation, there exists at least one real number root. Therefore, the equation $4x^3 - 3x = a$, of Wantzel's secondary principles has at least one real root solution. The real root solution can be formed a rational or irrational number. There is no other reason to determine the angle can be trisected or not.

Additionally, using ruler and compass only, to divide an angle into three equal angles is to divide an arc which is inscribed by the defined angle under the definition of angle measurement. The measurement of the angle is defined by the measure of an arc of a circle. To divide an arc is similar to dividing the circumference of a circle into n equal part. In other words, subdivide an angle $\measuredangle$ radians into n equable angles. We have achieved and learned to construct a regular n-gon, such as: n = 3, 4, 5, 6, 17, --- and so on for several hundred years. In the eighteenth century a young and bright

German mathematician Carl Friedrich Guss (1777-1855) proved the following theorem:

  If p is any prime number, then the reqular p-gon can be constructed using ruler and compass if and only if p can be represented in the form ;

  $P = 2^{2^m} + 1$, where m is a whole number;

Here, if m = 0, clearly, a circle can be divided into three equal arcs. It raises the very important question for us to rethink: " Can we divide a part of a circle into equal three parts?" For answering this question, let us study and investigate the 132 constructions and illustrations of Underwood Dudley's work. These methods of the drawings were to divide a given angle from the vertex directly into three equal angles including the constructing mehtod of the first greatest mathematician Archimedes (287-212 B.C.) without finding points on the arc which is inscribed by the given angle in the defined circle.

  In the past manner in dividing three equal major arcs on a circle was carried by the radius around the circles 6 times then carried out the procedure for the regular 6-gon, then, omitted every other point. These constructional skills were not precisely done. Theoretically, the entire total length of circumference of a circle with radius r is:

  C (circumference) = $2 \cdot \pi \cdot$ r Radian

  The number π = 3.1415 92655 8979323846 2643383279 50288419 71 6939937510 5820974944 5923078164 0628620899 8628034825$\cdots$, is going to an infinite place. Actually, the entire total length of circumference of a circle is not 6 times radius exactly. The precise construction can be found, if we use two basic drawing instruments skillfully.

First, we construct an equilateral triangle, then, we find
the center of the equilateral triangle by bisecting the three
interior angles. Finally, we draw the given circle at the center
of the equilateral triangle, as the result, the bisecting lines
of the three angles will divide three equal arcs of the circle
into six equal arcs including a semicircle into three equivalent

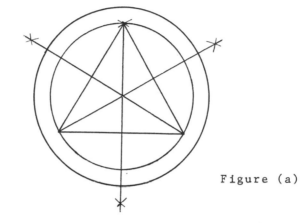

Figure (a)

Continuously, a right angle or a half right angle can be cut
off three equal arcs. Clearly, there exists more than one arc
of a circle which can be divided into three equal parts. Now we
are going to explore the generalization. First, we examine and
investigate some particular angles. According to the essential
Euclid's Elements the measurement of an angle is to be defined
between 0 and π radian. As we know, a right angle is $\pi/2$ radian,
if we are going to construct a right triangle with $\pi/6$, $\pi/3$, and
$\pi/2$ radian interior angles. Obviously, the angle $\pi/6$ radian is
one-third of $\pi/2$ radian, and the $\pi/3$ radian is two times of the
$\pi/6$ radian, if we bisect the angle $\pi/6$ radian into $\pi/12$ radian
which is a trisection of $\pi/4$ radian. Consequentially, the angles;
$\pi/2$, $\pi/4$, $\pi/8$, $\pi/16$, --- $\pi/2^n$ (n is a natural number) can be done
a trisecting an angle if the angle is big enough to employ a
straight-edge and compass. The right triangle is constructed
and discussed in the Chapter I.

For the generalization, if an angle α $(0<\alpha \leqslant \pi/3)$ is given then the angle α can be tripled by using two basic tools that are straight-edge and compass. Let the vertex of a given angle be O and two sides be $\overrightarrow{OX}$ and $\overrightarrow{OY}$, O_1 is an arbitrary point on the side $\overrightarrow{OX}$, also, a perpendicular line ℓ to $\overline{OO_1}$ is to bisect the segment $\overline{OO_1}$ at the point M and meet the side $\overrightarrow{OY}$ at a point named P, then let O_1 be a center and PO_1 be a radius and draw a circle to meet the line ℓ , the side $\overrightarrow{OY}$ of angle XOY at two points P and S respectively. After that, we draw $\overrightarrow{SO_1}$ and extend it to meet the circle O_1 at the point T. As a result, by the construction, the measurement of the angle OO_1T is exactly equal to three times of the measurement of the given angle XOY. The proof is started in the Chapter III. In the same manner, we can repeatly triple a given an acute angle XOY_1 $(0<m\angle XOY_1 \leqslant \pi/3)$ to be angle OO_1T_1 and so on. The basic tripling constructions are shown on the following Figure (b).

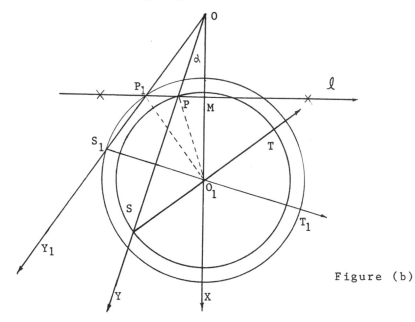

Figure (b)

According to the experimental constructions, the line ℓ and points O, M, and O_1 are situated without changing or moving. They are always kept constantly on the Euclidean plane[*].

---

* See page 62 and Appendix 111 & V.

On the other hand, we take the triple angle OO_1T to go backward to construct the original angle XOY grom the line ℓ and
the points O, M, O_1, and S. The point M is the middle point of
OO_1, the vertex O is the given angle XOY, the point O_1 is the
vertex of the tripled angle OO_1T. The line ℓ is perpendicular
to $\overline{OO_1}$ at the point M. The point S is the last defined and
intercepted on the the circle O_1 and opposited $\overrightarrow{O_1T}$. In another
construction, we let O_1 be a center with arbitrary radius and
draw a circle to meet the opposite side $\overrightarrow{O_1T}$ at S. After that,
we define the point M, the point O, and the line ℓ which is to
perpendicular and bisect $\overline{OO_1}$ and meet circle O_1 at the point P.
The following Figure (c) is the backward construction.

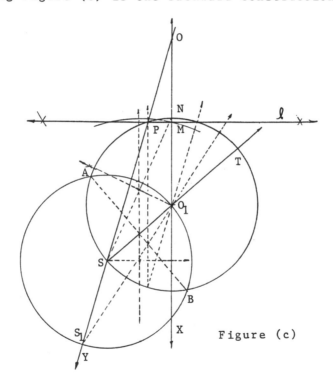

Figure (c)

Originally, the greatest Greek mathematician Archimedes in
the third century B.C., tried to define the point P first, but
he failed in definding the key point M. It has taken over
twenty five hundred years to define the point M. All entire
the theoretical and logical discussions are stated in the
Chapter III.

Here, we must base on the Euclidean Elements on an original and fundamental definition of plane geometry. A figure in the plane geometry is without a coordinated system to locate its position. A point canot be measured by its magnitute, but it is an element of a figure and has its location on a geometric plane. Generally, we can measure the metric distance between two distinct points. In mordern mathematics, a famous German mathematician Julius Wilhelm Richard Dekekind (1831-1916) had proposed an axiom that the points on a line can be put into one-to-one correspondence with the real numbers and the basic property for the points on a line:"If p, r are two different points, then there always exist infinitely many points that lie between p and r." Thus, the key point M at least exists between two points O and O_1. Also, the point P can be defined after the point M had been defined. The segment $\overline{PM}$ is a part or subset of line ℓ . Hence, PM always is perpendicular to $\overline{O_1M}$. Finally, we extended $\overline{O_1M}$ to O and let OM = MO_1. The point O is the vertex of the angle XOY.

Euclidean Geometry is concerned with the logical consequence of postulate system, when we say that a system of mathematical postulates is "true" we mean only, these postulates are free from contradiction. We continue developing mathematics topics in terms of these postulates. Here, we do not assert of comfirm in mathematics that anything actually exists in the physical world. However, there always exists this question: " Does the space we live in actually have the properties of the Euclidean space we study in Geometry?" The answer will be profound, we do not know well enough, furthermore, no mathematician ever expects to know. A geometric plane consists of infinitely many points. It will never be possible to determine whether the universe is finite or infinite in extent. Although, we have no assume that any point of the real world, actually, fits our geometrical postulates experience has shown us that much of our geometric theory is applicable to the world around us, and we can indeed use geometry to make very accurate figures or designings in the field of enginerring, architecture, and astronomy.

The new thinking of trisection theory is based on the logical consequence of postulate system from tripling an angle in the Euclidean Geometry. A logical set of postulates must be formed consistently; that is, that they do not lead to contradictory conclusions. Here, our postulate is: "If an angle is less than $\pi/6$, then it can be tripled." or "An angle always exists one one-third angle." or "One point must be related at least one point in the same geometric space." Those simple postulates contain no contradictions of logical reasoning. In the history of mathematics, a German mathematician David Hilbert (1862-1943) spent much of the latter portion of his life trying to prove the consistency of certain interesting mathematical systems. At about same time that he wrote his <u>Foundations of Geometry</u>, he showed that geometry is consistent if arithmetic is consistent. He had complished the proof that Geometry is consistent because Arithmetic is. All mathematicians believe that arithmetic is consistent. This belief is based upon basic operations of " + " " - ", " · ", " ÷ ", and " $\sqrt[n]{}$ " (n > 1 natural numbers.) They are not contradiction each other.

Finally, as we have stated on the first page, the trisected problem was discovered in the ancient Greek Geometry, It is more than two thousand five hundred years old. For solving the problem, mankind has yielded amazingly many new fruitful discoveries. Today, in addition to the new theory of solving trisection which is based on primarily geometry to study from tripling to trisecting an angle. Logically, there are eleven theorems which has been discussed and proved through an entire new theory. This theoretical study will be accepted or answered any comment or judgement in terms of the mathematical true. The three following figures (d), (e), and (f) are to test the accuracy of trisecting a sixty, ninety, and arbitrary degree angle by using advanced Autocad* program on the computerized drawing in terms of the new theory. As the result, the tested constructions are drawn and completed perfectly.

---

* see next page

The following figure was tested
by the Autocad program in the
advanced computer drawing. the
measurement of the angle is $\pi/3$.

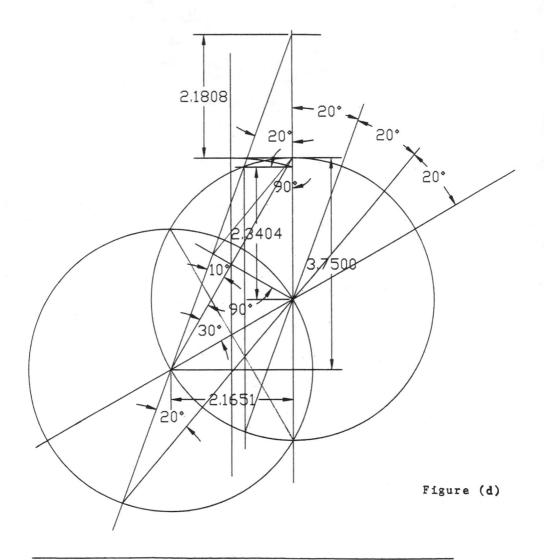

Figure (d)

---

* Autocad is defined by Computer Aid Design program
for architectural, engineering, mechanics drawings.
The program is produced by the Autodesk Company Inc.
San Rafael, California, U.S.A.

The following figure was tested
by the Autocad program in the
advanced computer drawing, the
measurement of the angle is π/2.

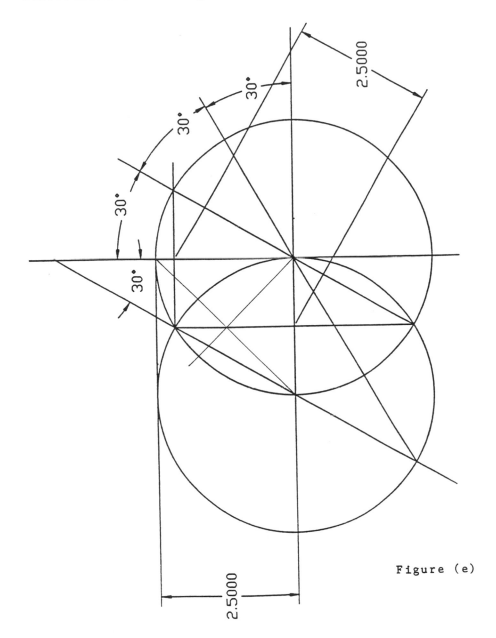

Figure (e)

The following figure was tested
by the Autocad program in the
advanced computer drawing, the
trisection angle is 50 degree
the decimal numbers are rounded
by the third digit place after
decimal point. The sum of the
three trisected angle is equal
to 50 degrees.

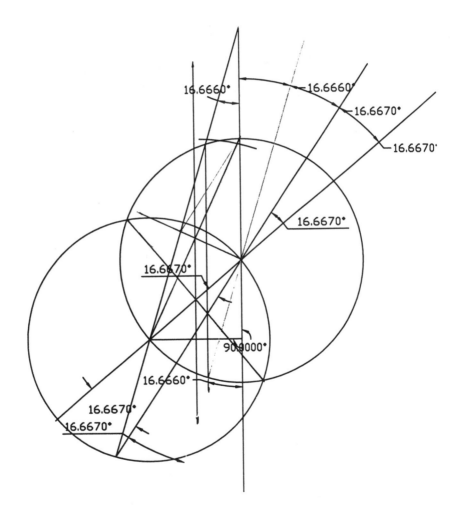

CHAPTER I

THE NATURE OF TRISECTION

ARCHIMEDES OF SYRACUSE
(ca. 287–212 B.C.)

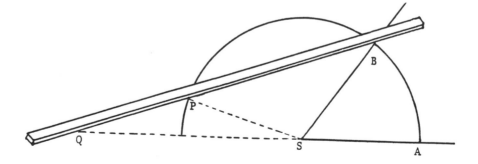

Chapter   I.   The Nature of Trisection

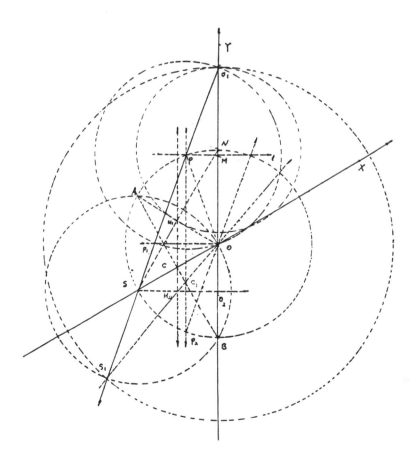

THE NATURE OF TRISECTION

I - I.   One of the Three Famous Problems

   In the history of mathematics, trisecting an angle was one of
the most difficult and vigorous problems in Euclidean Geometry.
The other two problems are: duplicating a cube and squaring a
circle.   The three unsolved problems now called three famous
problems because of their difficulty to be solved by using two
basic constructing instruments which are a compass and unmarked
straight-edge in the following Fig. I-1.   The three problems were
discovered in ancient times when Geometry was being invented.

(a)

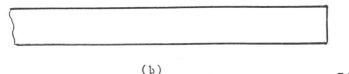

(b)

Fig.1-1.

   Geometry is one the greatest achievements of the human mind
by shaping the universe where we live on the earth. The knowing
knowledge of the shaping geometry facts and living needs goes
back before the beginning of recorded history.   The early ages
of Egyptians and Babylonians (4000-3000 B.C.) knew many beings
in fact practical and useful geometric relations.   The keenly
construction of the pyramids and the survey of the land along
with Nile river.   Continuously, the ancient Greeks collected,
the know geometric facts, discovered new ones, and arranged

them into a logical knowledge system. As we know, the word
geometry comes from two ancient Greek words: geo meaning is
"earth," and metrei meaning is "to measure," Therefore,
Geometry was thought of as "earth measurement."

   The knowledge of Geometry was continuing to organize and
discover. It took centuries just before, during, and after,
the period of the greatest political influence between 4th
and 5th centuries. At that time, the leading minds and ideas
were interested in developing the fundamental knowledge of
philosophy, astronomy, and geometry. Some of the geometers
had high achievements in searching intellectual activities
because of their strong influence on later men in the field
of philosophy, sciences or mathematics. One of these early
men was Thales of Miletus* (6th century B.C.) who was widely
considered one of the seven wise men of his time. Also, he
was considered the first Greek mathematician and the first
astronomer. He might have known how to prove the reasonings
behind geometric problems. In the developing knowledge of
geometry, the next important ancient Greek mathematician was
Pythagoras (late 6th century B.C.) who established a school
or brotherhood in Croton. His school was not only the great
achievement for the famous Pythagorean theorem (Fig. 1-2),
but also notable discoveries in music, astronomy, metaphysics
and arithmetic.

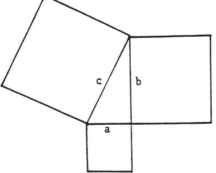

Fig. 1-2.

---

* Thales (ca 640-546? B.C.), Greek Philosopher born in Miletus,
The Random House Dictionary; Radom House, New York, 1967. P.1469.

In the 4th century B.C., one of the ancient Greek geometers was called Eudoxus of Cnidus during the time of Plato[*]. He had discovered "method of exhaustion" was one of the most useful methods of geometric construction which was based his general theory of proposition and golden section. An illustration of the way in which Eudoxus probably carried out his method of the exaustion. The proof of the areas of circles are to each other as squares on their diameters. Also, the proof, as it is given in Euclid's Elements X11.2 is probably that of Eudoxus is the following Fig. 1-3.

Fig. 1-3

The great philosopher Plato (4th century B.C.) of the ancient Greece founded the Academy. His student Eudemus of Rhodes, got credits "Hippocrates of Chios " with publishing in the earliest collection of Elements. At that time, many famous thinkers or mathematicians attempted to solve the famous problem in squaring of a circle was Anaxagoras[*] when he was in prison. There were no further details concerning the origin of the problem or the rules governing it. Anaxagoras died in 428 B.C., just one year before Plato's birth.

---

* Anaxgoras (500?-428 B.C.) a greek Philosopher, Carl B. Boyer: " A History of Mathematics." John Wily &Sons, Inc. New York, 1968 pp.63-4.

The nature of the squaring of a circle as shown the following Fig. 1-4. For attempting to square the circle, a theoretical question remaining a nice distinction between accuracy in approximation and exactitude in the mathematical thinkings.

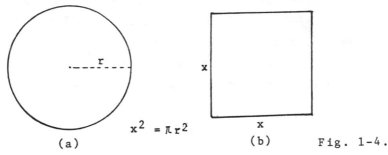

$$x^2 = \pi r^2$$

(a)   (b)   Fig. 1-4.

The second famous problem was to double the cubical altar of Apollo[*] at Delos which was asked for a method for skilling construction, with two basic geometrical instruments. Here, a cubic altar is going to have its volume doubled that of a given cublic altar. According to the legend, was the origin of the "duplication of the cube" problem, one that henceforth was usually referred to as the "Dolian Problem"--- given the edge of a cube, construct with compass and unmarked straight-edge alone the edge of a second cube having double the volume of the first in the following Fig. 1-5.

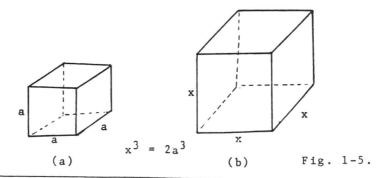

$$x^3 = 2a^3$$

(a)   (b)   Fig. 1-5.

---

* Apollo was the ancient Greek or Roman God of light healing, music, poetry, and manly beauty (a very handsome young man.)

The last famous problem was discovered after the duplicating
a cube. The third celebrated problem in Athens was: given an
arbitrary angles, construct by means of a compass and unmarked
straight-edge alone an angle one third as large as the given
angle as shown the following Fig. 1-6.

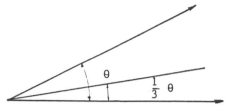

Fig. 1-6.

After Plato, the best known of the ancient Greek geometers
is Euclid, who organized the previous works of the Academy of
Plato, and wrote the Elements about 300 B.C. Euclid's work is
the most successful textbook in the history of mathemaics. As
we knew, the textbook was used throughout the world until well
into the present century. The Elements consist of 13 "books,"
the first six of which are concerned with plane geometry. The
others deal with arithmetic and solid geometry. In the history,
we do not know Euclid's birthplace, but we know that Euclid was
called to teach mathematics at the city of Alexandria which was
founded by Alexander in 332 B.C. Alexander was a supporter of
learning as was his successor Ptolemy I. It was Ptolemy was to
built the greatest library at the Alexandria, and was during his
rule that Euclid opened a school there. At that time, Ptolemy,
had learned Euclid's Elements.

There was no doubt, many able Greek mathematicians followed
Euclid, among whom were Archimedes (3rd century B.C.) and well
known Apollonius (3rd century B.C.) A problem commonly known
today as the "problem of Apollonius," that is: given three fixed
circles, to find a circle that touched them all.

The mathematician Archimedes who had perhaps the greatest mind of ancient times, did excellent works in the fields of mathematics, physics, engineering. He was the most successful mathematician to do the trisection problem in the ancient times. His solving process is showing in the Fig.1-7, he let the given angle Φ be trisected and vertex S be a center, then, he drew a circle of an arbitrary radius r, the circle S intersects two legs of the angle at A and B respectively. Then, he placed the paper wood strip on the figure in such way that it is passing through the point B and that an end point of the marked-off segment coincides with a point P on the circle, while the other end point coincides with a point Q (out of the circle S) of the extension of $\overrightarrow{AS}$ during PQ = PS. Then, the angle PQS is one third of the given angle Φ.

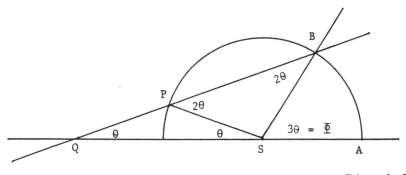

Fig. 1-7

In the year 300 A.D., the Greek mathematician Pappus proved and demonstrated in his ingenious masterwork: Συνα γωγα ι μ ‍ ‍ ατ ματιαι (Collections Mathematics). Pappus had formally proved the Archimedes' construction, since PS = PQ = r, thus, triangle PQS is an isossceles and angle PSQ is equal to θ. Also, since the triangle SPB is an isosceles too, consequently, m∠SBP = m∠SPB = 2θ. An isosceles triangle has two equal angles.

---

* m is a symbol of the measurement of an angle. Please see next section of the measurement of an angle.

Finally, the external angle o at S of the triangle SBQ is equal to the sum of the two nonajacent internal angles, that are angle SQB and angle SBQ. Therefore, we have:

$$m\angle ASB = m\angle PQS + m\angle PBS$$
$$\Phi = \theta + 2\theta$$
$$\Phi = 3\theta$$

or
$$\theta = \frac{1}{3}\Phi$$

As the proof above, Archimedes had tried to construct an angle equal to one third of the given angle indirectly, evidently, he did not directly trisect a given angle from vertex point, but; the problem of the trisection of an angle was the point P could not be defined by a compass or unmarked straight-edge. Pappus had tried to explain the possible construction; he first solved the problem:"find the locus of the vertex P of the triangle ABP with fixed base $\overline{AB}$, when the base angles A and B are to make up each other in proportion of 2 to 1." Because the method of the trisection was beyound the restriction of the basic and essence method of Euclidean geometry. Therefore, the Pappus' method was shown by historical interest.

By the historical review above, I found Pappus' method only looks at the point P on the Fig.1-7. He did not concern himself with the vertex Q of the finding angle. If we can defined the point Q first, then, we can define the point P easily. Because the point P is a intercepting point of the circle S and the line which is perpendicular and bisecting the segment QS. From those points remind me strongly to rethink this trisection problem in the two following foundamental and basic questions: " How do we define an angle?" and " How do we measure an angle with unit?" in the original of inventive or theorem which is derived by the definitions of its topic, postulates, and axioms in the entire knowledge body.

I - 2.   Angle and Angle Measurement

An angle was defined by the definition 8 and $\overset{*}{9}$ onthe Book I
of Euclid's Elements. The definition 8 is: " A plane angle is
the inclination to one another of two lines in a plane which
meet on another and do not lie in a straight line." According
to the original definition above, an angle is not considered a
straight angle, also the definition does not make clearly the
name of one vertex and two sides.   In the following Fig. 1-8,
let two lines be l_1 and l_2, if l_1 and l_2 are intersecting at
a point 0, then, they form two pair vertical angles.   We say
that ∠1 and ∠3 are a pair vertical angles, likewise ∠2 and ∠4
are a pair of vertical angles too.   As we know, two vertical
angles are congruent angles.

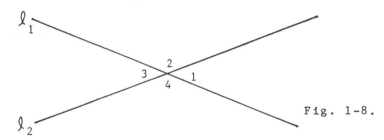

Fig. 1-8.

The definition 9 is: " And when the lines containing the
angle are straight the angle is called reclilineal."   As the
definition 9 is to modify the definition 8 for defing a basic
straight angle.   According to the definition, if two sides of
angle lie on one straight line, it might not form an angle
when two sides at same side of the vertex of an angle in the
following Fig. 1-9.

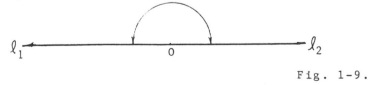

Fig. 1-9.

* Euclid: Elements, Vol. 1, Dover Publication, New York,
1956. pp.176-79.

A more precise definition of an angle is based on the morden
Set theory. An angle ($\angle$) is the union of two rays which have
the same end point. The two rays are called the sides of the
angle and their common end point is called vertex.

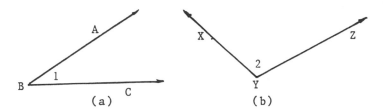

(a)   (b)   Fig. 1-10

The defined angle at the left (a) of Fig. 1-10 above, is the
union of rays $\overrightarrow{BC}$ and $\overrightarrow{BA}$. Its vertex is B and its sides are rays
$\overrightarrow{BC}$ and $\overrightarrow{BA}$. It can be represented in symbols by $\angle ABC$ or $\angle$ CBA or
$\angle 1$. The angle at the right (b) of the Fig. 1-10 above, is the
union of rays $\overrightarrow{YZ}$ and $\overrightarrow{YX}$. Its vertex is at Y and its sides are
rays $\overrightarrow{YZ}$ and $\overrightarrow{YX}$. Also, it can be represented symbols by $\angle XYZ$ or
$\angle ZYX$ or $\angle 2$. Particulary, here, when three letters are used to
represent an angle, the letter at the vertex is always placed,
between the two other letters.

Next, we are going to discuss the measurement of an arbitrary
angle. In general, as we defined the measuring system in length
of segment or distance of two defferent locations. We normally,
choose units, such as; mm, cm, m, km,---etc., or inch, foot, and
mile,---etc.. Similarly, to measure angle, we need to define a
standard unit. There are: the degree, the radian, and the mil.
A mil is very small unit, approximately 0.56 of degree. It is
not commonly used

The degree system is more popular in the measuring an angle.
The main question is: How do we define one degree? Originally
we start to draw an unit circle, then, we precisely divide the
circle into 360 equal slices from the center, each equal slice
is called one degree. Therefore, the total degrees of central
angle are 360.

Traditionally, one degree can be represented a symbol 1°. For a more small unit, we divide one degree into 60 slices. each slice is called a minute which can be represented 1'. After that, one minute can be divided into 60 equal slices, each slice is called a second. One second can be symbolized 1". It is exactly similar to the time units. The following Fig. 1-11 is defining the degree units.

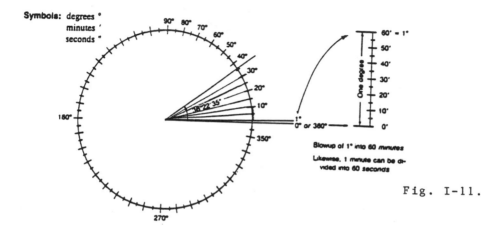

Fig. I-11.

Finally, We have defined a radian system for measuring the length of arc or circumference of a circle. The Greek letter π[*] is used to denote the irrational number which is the ratio of the length of circumference (C) of any circle to its length of diameter (d). Since π is irrational, it canot be indicated a repeating or terminating decimal. The ancient Babylonians and Egyptians knew more about it than its mere existence, they had also found its approximate value by about the year 2000 B.C. The Babylonians had nearly arrived at the value $\pi = 3.125$ and the Egyptians at value $\pi = 4(8/9)^2$. For computing the closed value, we can calculate into several thousands digital places after decimal point by using computers in the 20th century.

---

[*] Peter Beckmann: A history of π, The Golem Press, 1971, P.11.

The number π is not only called an irrational number but also called a transcendental number. In mathematics history, the number π is called Archimedean number.

In theoretical mathematics, a radian system is used widely and intensively, If we construct a unit circle, the length of circumference of the circle is 2π. But, if the radius of a circle is not equal to one unit, then, the entire length of circumference of the circle is equal to $2\cdot\pi\cdot r$ (r = radius.) However, the degree system with 360 degrees marks off the all entire circle. Let us define a function M for exchanging two measuring systems between radian and degree. Also, let domain be D = [0, 360), and codomain be R = [0, 2π). Cleary, the function M is one to one corresponding relationship between D and R. The domain D is a rational number, but, the codomain is an irrational number. As the function, we have:

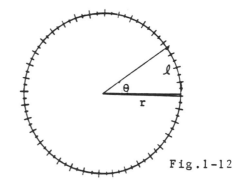

$$M \; (\; x \;) \; : \; D \longrightarrow R$$
$$M \; (\; 0 \;) \; = \; 0 \quad radian$$
$$M \; (90 \;) \; = \; \frac{\pi}{2} \quad radian$$
$$M \; (180) \; = \; \pi \quad radian$$

or $\quad 180 \; x \; 1° \; = \; \pi \quad radian$

$$1° \doteqdot 0.017145 \; radian$$

And, 1 radian $= \dfrac{180°}{\pi} \doteqdot 57.295778°$

Fig.1-12

As we have defined two major systems for measuring an angle, we know what is meant by arc length; it will be easy to scale or measure an angle when drawing a circle with its center at the vertex. The minor arc is inscribed by two sides of the defined angle; the length of the minor arc is the measurement of the defined angle. If the vertex of a defined angle is a center of a circle, then, the angle is called a central angle which will be discussed in the Chapter II. Here, suppose that ℓ is the length of inscribed arc with a circle of radius r, the measurement of the angle is θ radian, then ℓ = r x θ radian.

In the previous discussion, we have in effect been measuring angles by measuring arcs. In fact, we have expressed the arc as a fraction of $\pi \cdot r/2$ (r is a radius of a circle,) that is as a fraction of one quarter of the circumference. Actually, the $\pi \cdot r/2$ is not a rational fraction because the number π is not a general algebric number. Therefore, for rethinking trisection problem, it is very important for us to examine and verify a relationship between an arc and its radius in a defined circle.

For finding a mapping relationship, suppose a circle with a radius r (r > 0) on a XY-plane coordinated system, also, if the center of the circle is defined at the original point (0,0), then the equation and its figure (Fig. 1-13) are:

$$X^2 + Y^2 = r^2 \quad (r > 0)$$

$$Y^2 = r^2 - X^2$$

$$Y = \sqrt{r^2 - X^2} \quad (-r \leq X \leq r) \text{ --- (1)}$$

$$Y = \sqrt{r^2 - X^2} \quad (0 \leq X \leq r) \text{ --- (2)}$$

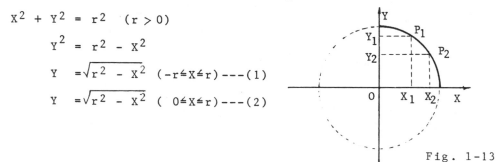

Fig. 1-13

The equation (1) is represented by the entire circle, but the equation (2) is only describing a quarter arc of the circle at the first quadrant of the coordinated plane. If we let function be: $\lambda : X \longrightarrow \sqrt{r^2 - X^2} \ (0 \leq X \leq r)$; then the function must be a one to one function.

For proving function λ is an one to one function, we take X_1 and X_2 ($X_1 \neq X_2$) then, we have $\lambda(X_1) = \sqrt{r^2 - X_1^2} \ (0 \leq X_1 \leq r)$, and $\lambda(X_2) = \sqrt{r^2 - X_2^2} \ (0 \leq X_2 \leq r)$. Since $X_1 \neq X_2$, obviously, we have, $\lambda(X_1) \neq \lambda(X_2)$. Hence, by the definition, the function is an one to one function. Also, its inverse function $\lambda^{-1}(X)$ is the same as $\lambda(X)$. The length of arc is not equal to the length of the radius, but the points between arc and radius are one to one correspondence to each other.

I - 3. Ambiguity Proof

To construct a geometric figure, is a classical game in the
ancient greece. The Greek philosopher and mathematician plato
and geometers have made a clear distinction between a drawing
and a construction. A drawing is a representation made on a
slates or stone or wall or wood using a craze or charcoal or
chalk or knife with any suitable drawing instruments such as
a protructor, T-square, marked ruler, compass, or straight-edge.
A geometric construction was a game to be played by using simple
equipment such as a compass and an unmarked straight-edge which
have been described on the Fig.1-1.

The basic rules govering the use of these two dull permissible
instruments are as follow:

1. An unmarked straight-edge can be used only to draw
   a straight line through two given distinct points.
2. A compass can be used to mark off equal segments,
   construct a circle, and an arcs of circles.

After the three famous problems were discovered the ancient
Greeks, many professional mathematicians and amateur thinkers
became interested by solving the problems. These inquisitive
mathematicians or thinkers devoted their talent to discovering
new apparatus and theorems for their solutions. During the time
for solving the very difficult and vigor problems, new fields
of mathematics were to be grown and developed, such as; algebra
coordinate geometry, trigonometry, calculus, and modern algebra
century after century, But the three famous problems are still
unsolved in the history of mathematics.

Until the year 1837 a young German mathematician Pierre
Laurent Wantzel (1814-1849) proved that there is no unmarked
ruler and a compass construction that can exactly divide an
arbitrary angle into three equal parts.

Wantzel's proof of this theorem is based upon the following two principles as the following statements:

1. A given angle α can be constructed if and only if it is possible to construct a line segment of length $\cos \alpha$; and therefore from what we have only limited to use two basic instructional tools, that an angle α can be drawn, if and only if the number $\cos \alpha$ can be constructed using the four operations of arithmetic, +, -, ', $\div$, and square root ($\sqrt{}$) operation.

2. From the well-known trigonometric identity

   $\text{Cos}3\beta = 4\cos^3\beta - 3\cos\beta$.

   It follows that for any angle α we have (putting $\alpha/3 = \beta$) $\cos\alpha = 4\cos^3(\alpha/3) - 3\cos(\alpha/3)$. Now let α be any given an angle and define a $= \cos\alpha$ and $X = \cos(\alpha/3)$. we can trisect the angle if and only if the number $\cos(\alpha/3)$ is able to be constructed, and therefore we can trisect the angle if and only if the roots of the algebraic equation:

   $4x^3 - 3x = a$ ----- ------ - (1)

   are constuctible numbers.

According to the first principle of Wantzel's statement, if an angle   is given, we can construct an angle α to an central angle with a unit circle in terms of the definition of $\cos$ =0A as following Fig. 1-14.

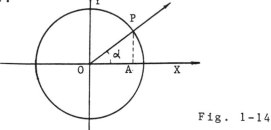

Fig. 1-14

Inversely, if the number of $\cos\alpha$ is given, then let us be able to construct an angle α . The construction is impossible because the given is a number, it is not a point set. By the definition of function, geometric measurement is a function, but it is not always a one to one function. Consequently, a measurement function is not always existing as an inverse.

Second, the geometric construction is not binary operation
is arithmetic or algebric, +, -, ·, ÷, and squaring root
operations. Because the Euclidean geometry is not only to
study the measurement of a segment, an angle, an arc, or
an area of a figure, but also to discuss the relationship
and properties about points, segments, angles, lines, arcs,
circles, and different shape of figures in the geometric
space. For examples, line ℓ_1 + line ℓ_2 are equal what? Two
lines are not always intercepting each other to form an
angle or one pair vertical angles. Thus, the addition for
two lines is no meaning. One angle A + one angle B are,
equal what? If the two angles are not at the same plane,
then, they are no meaning to discuss. The operation of a
construction canot be followed by the operation of modern
algebric field. Furthermore, geometric construction is not
only able to construct a square root segment, but also a
fourth root segment, eight root segment, ---, and so on.

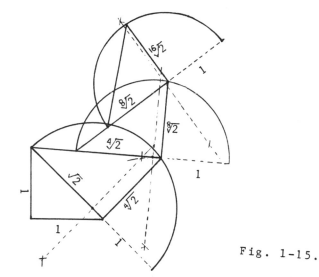

Fig. 1-15.

The Fig.1-15 above, is to construct a segment of $\sqrt{2}$, $\sqrt[4]{2}$,
$\sqrt[8]{2}$, ---, and so on. The question is: " Can we construct a
segment equal to $\sqrt[3]{2}$ between $\sqrt{2}$ and $\sqrt[4]{2}$?"

Third, in the second principle of Wantzel's proof, we can
trisect the angle if and only if the roots of the algebraic
equation: $4x^3 - 3x = a$ are constructible numbers. Wantzel's
proof the equation does not have rational solutions, hence,
an angle canot be trisected. For this reason, we can argue
that Wantzel's proof was ambiguous. Why?

The cubic equation had been solved into an algebric formula
by an Italian physician and mathematician Jerome Cardan (1501-
1576) in the sixteen century. The three roots of the cubic
equation always have one real number root and a pair conjugated
complex numbers or three real numbers. In other words, the
solution of a cubic equation with the coefficients over real
numbers, has at least one real number root or three different
real number roots, The roots of the cubic equation might be
rational or irrational numbers. Therefore, there exist in the
general solution. If we take the equation into a polynomial
function, then the graph of the function will be demonstrated
clearly by the roots relationship with x-axis of the Euclidean
coordinated plane.

As we have followed Wantzel's proof, the variable x is cosine
function ($x = \cos\alpha/3$). The variable is a value of $\cos\alpha/3$, it
is not direct to the inscribed arc of the angle. Now, if we
let the function be:
$$f(x) = 4x^3 - 3x - a, \quad a \text{ is a constant number,}$$

Obviously, we can investigate the graph of the function f in
terms of the constant number a.

Case 1. If $a = 0$; then, we have:

$$f(x) = 4x^3 - 3x ;$$

Or
$$f(x) = x(4x^2 - 3);$$

After that, we let $f(x) = 0$ for finding the solutions of the
equation $f(x) = 0$ or $x(4x2 - 3) = 0$;

And,
$$x = 0 \text{ or } 4x^2 - 3 = 0 ;$$
$$x = 0 \text{ or } x = \frac{\sqrt{3}}{2} \text{ or } x = \frac{-\sqrt{3}}{2} ;$$

The equation $f(x) = 0$ has three different real number roots
o, $\frac{\sqrt{3}}{2}$, and $\frac{-\sqrt{3}}{2}$.   Two irrational roots of the equation are
constructible square roots.   The graph of the cubic function
is intercepting three points with x-axis on a   coordinated
plane as shown the following Fig. 1-16.

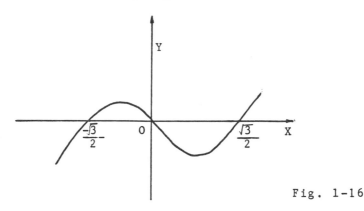

Fig. 1-16

The three roots of the graph above, bring us to the that one
solution during $\alpha = 90$ in terms of the definition of an angle
in Euclidean geometry. Now, we are going to construct an angle
which is one third of a right angle. Let angle XOY be a right
angle, then construct its one third angle by using a compass
and unmarked straight-edge.

Construction:

    1. Let O be a center with an arbitrary radius r, then, we
       draw a circle O to intercept ray $\overrightarrow{OY}$ at the point N and
       the opposite ray $\overrightarrow{OX}$ at the point S.

    2. Let the point S be a center with radius r and draw an
       equal circle O.   Then, named two points A and B, are
       the intercepting points of the circle O and circle S

    3. Draw a ray $\overrightarrow{SB}$ to meet the ray $\overrightarrow{OY}$ at the point O_1 with
       the circle S at the point S_1.

    4. The angle SO_1O is one third of the given right angle.

We have used two basic constructing instruments and followed
four drawing steps as shown the following Fig. 1-17.

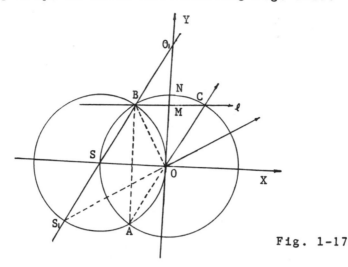

Fig. 1-17

The construction above, is to construct a specializing right
triangle in which the measurements of three interior angles
are: 30, 60, and 90 degrees. Especially, the ratio of the two
interior angles is 1 : 2. The construction has been completed,
we must answer the question:" Is the construction correct? "

Proof: On the Fig. 1-17, if we connect segment $\overline{OB}$, then the
triangle OSB is an equilateral triangle. Hence, we have the
angle BSO, 60 degrees. But, the given angle XOY is a right angle,
Thus, we have the angle SOO_1 measuring a right angle too. As a
result, the measurement of the angle SO_1O is 30 degrees in the
constructed right triangle SOO_1. Hence, the angle SO_1O is one
third of the right angle XOY.

For the trisection of a right angle, if we draw a line ℓ and
through the point B to perpendicular the ray $\overrightarrow{OY}$, then, we name
the point M is the intercepting point of the line ℓ and ray $\overrightarrow{OY}$.
Obviously, M is the middle point of the segment $\overline{OO_1}$. If we name
point C as the line ℓ intercept circle O on the arc of the given
angle XOY, as the result, the point C is the trisecting point of
inscribed arc of the given right angle XOY.

Case 2.   If a = 0, then we differentiate the function f(x):

$$f\ (x) = 4x^3 - 3x - a$$

Or, $f'\ (x) = 12x^2 - 3$

And, $f''(x) = 24x$

Let $f'\ (x) = 0;\quad 12x^2 - 3 = 0\ ;$

The two roots are: $x = \dfrac{-1}{2}$ and $x = \dfrac{1}{2}$

Obviously, there are two critical points at $x = \dfrac{-1}{2}$ and $x = \dfrac{1}{2}$.   One is the maxima point, the other one is a minima point on the curve of $y = f(x) = 4x^3 - 3x - a$.

Futhermore, we let $f''(x) = 0$, then, we have:

$$24x = 0$$

Or $x = 0$,   Clearly, at the point $x = 0$, there exists an inflection point between $x = \dfrac{-1}{2}$ and $x = \dfrac{1}{2}$.   The graph of the function $y = f(x)$ intercepts at least one point with x-axis. The possible graph as following (a) and (b) of the Fig. 1-18.

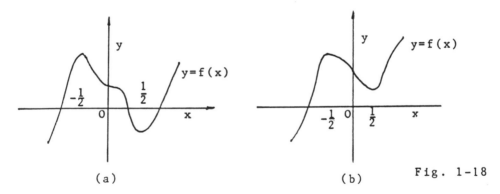

(a)   (b)   Fig. 1-18

From the two graphs above, each graph exists one inflection point between two critical points.   Therefore the curve of the function $y = f(x)$ intercepts x-axis at least one point or three points.   However, the equation $f(x) = 0$ always exists at least one real root solution.

After the investigation, the graphs of the defined function:
$f(x) = 4x^3 - 3x - a$ are showing to intercept x-axis one point or
three points which are to interpret the solutions of equation:
$4x^3 - 3x - a = 0$ has at least one or three real roots in terms
of the fundamental theorem of algebra. If we let $x = \cos \alpha/3$,
then, it is over real numbers including rational or irrational
numbers. As a result, how can the cubic equation be solved by
only rational roots? Of course, the solutions of the equation
can have real or imaginary numbers. The real solutions can be
rational or irrational numbers. As the result, Wantzel's proof
contradicts his statements of the assumption. Obviously, the
Wantzel's proof is not completely valid in the original theory
of Euclidean geometry, because Wantzel empolyed a new algebraic
equation without using geometric figures.

On the other hand, the trisection problem is different from
the other two famous problems. They are going to construct equal
areas between a circle and a square or double a given cubic in
volume. For the trisection problem, if we look back to the Fig.
1-8, you will find Achimedes' construction tried to define a key
point for solving the trisection. He did not completely define
the key point. Thus, how to define the key point clearly became
the main issue for solving the trisection. The key point on the
arc must be relative to another point in the same point-set-space
of the Euclidean geometry.

It is many years since Wantzel's proof that is impossible to
trisect an angle into three angles by using an unmarked straight
edge and a compass. But the reminded question is: " Why have so
many mathematicians devoted their valuable and useful time into
solving the trisection problem?" Recently, there was a book " A
Buget of Trisection." The book was organized and published by
professor Underwood Dudly in the year 1989. The book described
titled and listed 132 works on the trisection problem. I belive
the thinkers must have discovered Wantzel's proof is not valid.

CHAPTER II

A NEW APPROACH TO THE TRISECTION

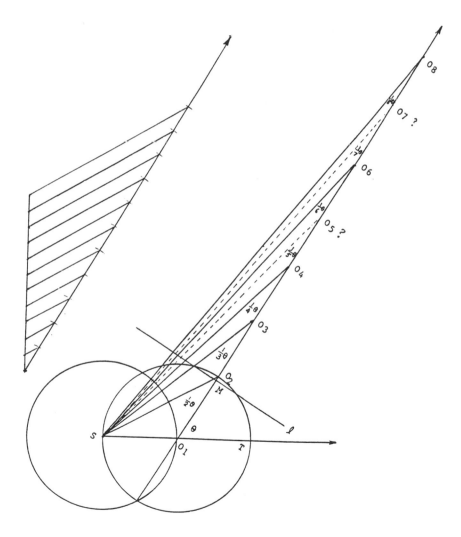

Chapter　　II.　A New Approach to the Trisection

II-1.　A New Assumption in the Euclid's Elements

II-2.　Arc, Central and Inscribed Angle

II-3.　Doubling and Bisecting an Angle

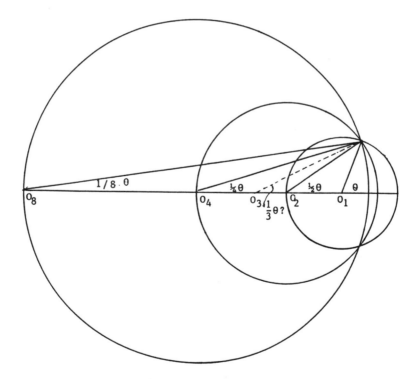

A NEW APPROACH TO THE TRISECTION

11 - 1   A New Assumption in Euclid's Elements

In the last section of the Chapter I, we have discussed
the ambiquity of Wantzel's proof. His assumption were not
based on the transcendental number pi (π) for measuring
an angle, he proved that an angle could not be trisected
by using numerical quantity under an algebric operation
without a geometric figure and an applying a theorem in
Euclidean Geometry. He had ignored the basic formation of
the entire knowledge of geometry.

Geometry is based on assumptions, that are combined by
a set of undefined terms and a set of statements about them.
Those statements are supposed to be true without proving.
Thus, the statements are called postulates or axioms. From
these assumptions, we define new terms or words and prove
other statements by logical process of deduction, all the
statements that are proved from the postulates by deduction
are called theorems or propositions.

Euclid's Elements made his set of axioms and postulates[*]
as small as possible by following the ancient Greek idea of
simplicity of form. The ancient Greek tried to reduce every
geometric construction to its most basic ingredients. As a
result, an additional important idea to the study of Euclidean
geometry is that of geometric construction since the time
Plato (4th century B.C.) in the ancient Greek's game. At
that time, a formal geometric construction allows only two
tools; an unmarked straight-edge and a compass as we have
stated and mentioned on the Fig. 1-1.

---

* see BOOK 1, Vol.1 of the Euclid's Elements, over Publications,
  Inc. New York, 1956 pp. 154-5.

In the Book 1 of the Euclid's Elements, there are five axioms and five postulates:

Axioms.

1. Things which are equal to the same thing are also equal to one another.

2. If equals be added to equals, the wholes are equal

3. If equals be subtracted from equals, the remainders are equal.

4. Things which coincide with one another are equal to one another.

5. The whole is greater than the part.

Postulates.

1. A line can be draw from any point to any other point. In the Euclidean Geometry a line is a straight line.

2. A finite line segment can be extended to a line of any length.

3. A circle can be draw with any center and at any distance from that center.

4. All right angles are equal to one another.

5. Through a given point not on a given line only one line can be drawn parallel to the given line. --- It is the parallel postulate; parallel lines are lines in the same plane that never intersect or meet, no matter how far they are extended.

According to the axioms and postulates above, the postulates are dealing with geometric construction in a point set.or a figure. The axioms are to operate quantities with arithmetic properties. Therefore, the entire knowledge systems are formed by a quantitative measurement and nonquantitative point set.

From the postulates, we can use an unmarked straight-edge and a compass and develope the following basic construction:

. a congruent line segment

. a perpendicular to a line at a point on the line

. a perpendicular to a line from a point not on the line

. a perpendicular bisector of a line segment

. a congruent angle

. a parallel line

. the division of a line segment into any number of equal parts

. an angle bisector

. an arc bisector

. a tangent to a circle at the point on the circle

. a tangent to a circle from a point outside of the circle

. a circle circumscribed about a triangle

. a circle inscribed in a triangle

Each of these basic constructions can be proved or justified by relevant theorems. However, we can develop an intricate or complex construction from some basic constructions. In addition, Euclid demonstrated that a segment of any line segment can be reproduced on any other segment because the transfer can be accomplished with a compass.

We have learned that a set of assumptions is a core of the formation. Therefore, for solving the trisection problem, we are adding one postulate, that is: "If a point exists in the Euclidean space, the point must be related to at least one other point in the its space." Because the key to solve the trisection problem is how to locate exactly the point P on an arc in the Fig. 1-8.

11 - 2   Arc, Central and Inscribed Angle

In the Postulate 3*of the Euclid Elements.   A circle is
defined:"to describe a circle with any center and distance."
The distance means radius is a segment from the center to a
point on the circle.   In modern words or symbols, a circle
can be defined by the set all points on a plane, that is a
given distance from a fixed point on the plane.   Let O be
the center   and OA be a radius of the circle, the points B,
C, are on the circle as the following Fig. 11-1.

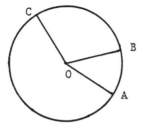

Fig. 11-1

By definition of a circle, OA = OB = OC = --- and so on.
Now, OA, OB, and OC are called radii.   Naturally, radii of
a circle and equal.   Inversely, if we define two circles
they have equal radii , then the two circles are naturally
called two equal circles.   On the following two circles,
are equal circles because they have two equal radii; OP
and OQ are congruent to each other.

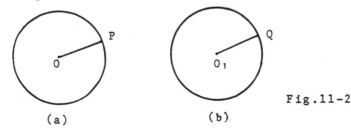

Fig.11-2

(a)   (b)

On a defined circle, if two end points of a segment are
on the circle, then the segment is called a chord. Thus, a
chord always divides the defined circle into two regions.

---

* On the Postulate 3 of the Book 1 of the Elements
  " To describe a circle with any center and distance." (p. 199)

If a chord is passing through the center of the defined circle, then the chord is called a diameter. Consequently, the length of a diameter always is equal to two times of the length of a radius of the same circle or equal circles. The following Fig. 11-3 are showing two equal circles.

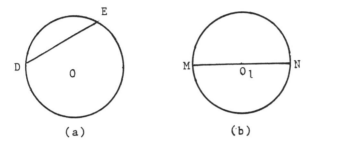

Fig. 11-3

(a)   (b)

Previously, we have discussed a definition of an angle. An angle is defined by three elements; that are one vertex and two sides. Here, we are going to define the relations between an angle and a circle. If the vertex of an angle is the center of a defined circle, then the angle is called a central angle, In the following Fig. 11-4 , the angle AOB is a central angle, if the points A and B are on the circle. All points of the circle which lie in the interior of the central angle AOB is called a minor arc AB or $\overset{\frown}{AB}$. Evidently, on the other hand, all points of the circle which lie in the exterior of the central angle AOB is called a major arc. Here, normally, we named a point X to call a major arc AXB or $\overset{\frown}{AXB}$.

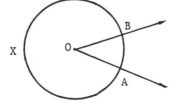

Fig. 11-4

By the definition of the measurement of a minor arc is defined by its central angle in the previous chapter, the unit of a basic measurement of arc is radian.

In the same circle or in equal circles, arcs have equal measurement are called equal arcs. If two central angles are equal, then, their minor arcs are equal to each other. Inversely, if two minor arcs are equal, then their central angle are equal to each other too. As a result, we easily obtain the following Theorem 11-1 in terms of the definition of the measurement of an angle.

Theorem 11-1 If only if two minor arcs are equal then, their central angles are equal in the same circle or equal circles.

The vertex of an angle can be located any place on a plane. If the vertex of an angle lies on the circle, then the angle is called an inscribed angle or is called to be inscribed in the circle. If angle ABC, the vertex B lies on the circle O the sides of the angle ABC contain chords $\overline{BA}$ and $\overline{BC}$ then the angle ABC is an inscribed angle is said to be inscribed into the circle O. The angle ABC is also said to be inscribed in arc ABC, In other words, angle ABC intercepts or cut off the arc AC.as the following Fig.11-5.

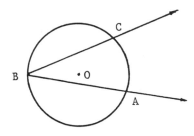

Fig. 11-5

Here, it is very important for us to focus on the vertex of an angle where it is located with a circle. There exists measuring relationship between central and inscribed angles.

According to the definition of an inscribed angle, let three
points; A, B, and C be on a circle O, and let angle BOC be a
central angle. Also, let the angle BAC be an inscribed angle
at the same circumference or arc BC as base. After that, we
connect $\overline{AO}$ and extend it to meet a point E on the circle O.

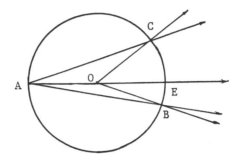

Fig.11-6

From the Fig. 11-6 above, since segments $\overline{OA}$ and $\overline{OB}$ are equal
radii, thus, the angle OAB is equal to the angle OBA. Clearly,
the angle BOE is one of an exterior angle of the triangle AOB.
By the relationship between an exterior angle and its two acute
remote angles,[*] therefore, we have:

    m∠BOE = m∠OAB + m∠OBA ; m is the measurement of an angle.

or m∠BOE = 2m∠OAB = 2m∠OBA -----------------(1)

   At the same reason, we also have:

    m∠COE = m∠OAC + m∠OCA

or m∠COE = 2m∠OAC = 2m∠OCA -----------------(2)

   If we add both sides of (1) and (2), then, we obtain:

    m∠BOC = 2m∠BAC

Consequently, we have the statement: the measurement of the
whole angle BOC is double the measurement of the whole angle
BAC. Obviously, we have the following theorem.

   Theorem II-2[**]: The measurement of an inscribed angle is
one-half the measurement of its intercepted arc.

---

* On the proposition 32 of the Book 1 of the Elements (p.316)

** On the proposition 20 of the Book 111 of the Elements (p.46)

11 - 3   Doubling and Bisecting an Angle

In the Theorem 11-2, we see the relationship between central
and inscribed angles in intercepting the same arc and equal arc
on the same circle or equal circle. Actually, in other words,
the measurement of a central angle is double the measurement of
an inscribed angle in the same intercepted arc on the same circle.
Now, we can apply the theorem 11-2 to construct a doubling angle.
First, let angle ABC be an acute angle, the point B be a vertex
and the point A be a point on one side of the angle ABC. Second,
Let AB be a diameter, using A and B as center, construct arcs on
both sides of $\overline{AB}$. These arcs will intersect in the points P_1 and P_2,
then, draw $\overline{P_1P_2}$ and intercept $\overline{AB}$ at the point O. Obiviously, the
point O is the midpoint of $\overline{AB}$. Then, let the point O be a center
and $\dot{O}A$ or OB be a radius and draw a circle. Finally, construct
a ray $\overrightarrow{OC}$. As the result m∠BOC = 2m∠BAC because the central angle
BAC and the inscribed angle are intercepting a same arc $\overparen{BC}$. Here.
the point C is an intercepting point of ray $\overrightarrow{AC}$ and circle O. The
following Fig.11-7 is the doubling construction.

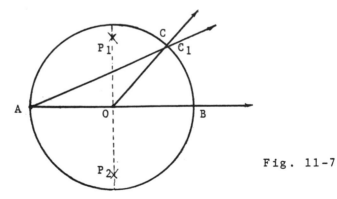

Fig. 11-7

Continuously. using the same manner, we can construct a double
angle BOC. If the angle BOC is still an acute angle, then let
$\overline{OB}$ be a diameter, using O and B as center, construct arcs on the
both sides of $\overline{OB}$. The arcs will intercept $\overline{OB}$ at the point O_1 .

Consequently, the point O_1 is the midpoint of $\overline{OB}$. Then, we let the point O_1 be a center and OO_1 or O_1B be a radius and draw a circle, the circle O_1 intercepts the ray $\overrightarrow{OC_1}$ at the point C_2. By the Theorem 11-2, we obtain $m\angle BO_1C_2 = 2m\angle BAC$. If the angle BO_1C_2 is still an acute angle, then, then, we can continue to double the angle BO_1C_2 --- and so on. The following Fig. 11-8 is showing the doubling construction.

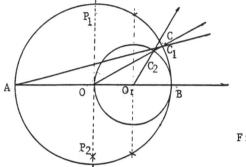

Fig. 11-8

For generalization, the question is:" Can we construct two points C_1 and C_2 at the same point C on the Fig. 11-8 above?". On the other hand, we let OC_1 be a diameter, then bisect $\overline{OC_1}$, its bisecting line meets $\overline{OB}$ at O_1. After that, we let O_1 be a center and O_1C_1 or O_1B be a radius and draw a perfect circle. The circle $\odot O_1$ must pass through the point C_1 or C.

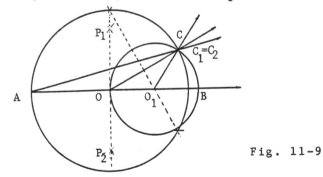

Fig. 11-9

manner, if the angle BO_1C_1 is still an acute angle, then we can continue to double the angle BO_1C_1 --- and so on. The Fig. 11-9 above, is very helpful for us to find a generalized method for doubling an angle.

In the previous paragraphs, we have discussed how to double
an acute angle by using the relationship between the central
and inscribed angles. As we knew, since the days of the Greek
philosopher and mathematician Plato (ca.380 B.C.) men had keenly
bisected from a given angle. In the other words, a historical
method of construction is to divide a given angle into two equal
angles, at the same vertex of the given angle. Let us examine
or investigate carefully to the traditional bisecting method.

   suppose a given angle ABC, now, we are going to construct a
ray which bisects a given angle ABC. First, let vertex B be a
center with any arbitrary or convenient radius, and construct
an arc intersecting the side $\overrightarrow{BA}$ at P and the side $\overrightarrow{BC}$ at Q.,
Second, with the points P and Q as centers and a radius greater
than ½PQ, construct arcs intersecting at the point R. Finally,
draw ray $\overrightarrow{BR}$, the ray $\overrightarrow{BR}$ is to bisect the given angle ABC.

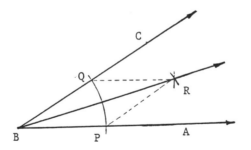

Fig. 11-9

   In this construction , we can easily prove that the angle
RBP and angle RBQ are congurent to each other. Since $\overline{BP}$ and
$\overline{BQ}$ are drawn, they are congruent segment. For the same reason,
$\overline{PR}$ and $\overline{QR}$ are congruent segment too. And the segment $\overline{BR}$ is
congruent itself. Thus, the triangle RBP and triangle RBQ are
congruent to each other. Therefore, the angle RBP and angle
RBQ are congruent angles, the ray $\overrightarrow{BR}$ is a bisector of the given
Angle ABC.

Euclid stated in the Elements, two congruent angles are the
measurement of equal arc in the same circle or congruent circles,
the two congruent angles are not necessary to have same vertex or
common sides.   For bisecting an angle, we can construct an angle
its measurement equal to one-half of the measurement of a given
angle by applying the Theorem 11-2. Let a given angle be angle
XOY.   First, let O be a center with arbitrary radius and draw a
circle intersecting the side $\overrightarrow{OX}$ at A.   The side $\overrightarrow{OY}$ at B, and the
opposite side $\overrightarrow{OX}$ at C.   Second, draw ray $\overrightarrow{CA}$ and ray $\overrightarrow{CB}$.   As the
construction, the inscribed angle ACB is one-half measurement of
the given angle XOY or angle AOB.   The construction of bisecting
angle XOY as shown in the following Fig. 11-11

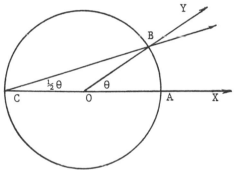

Fig.11-11

According to the construction above, clearly, the angle AOB
or angle XOY is a central angle and the angle ACB or angle BCA
is an inscribed angle in the same circle O.   Two angles have
intercepted at the same minor arc AB of the circle O.   By the
Theorem 11-2, we have:

$m\angle ACB = \frac{1}{2}m\angle AOB$

or $m\angle ACB = \frac{1}{2}m\angle XOY$

If let $m\angle AOB = \theta$

then, $m\angle ACB = \frac{1}{2}\theta$

Continuously, let the point C be a center, the segment $\overline{CB}$ be a radius, then, we draw a circle D. The circle D intercepts the ray $\overrightarrow{DA}$ at A_1 or angle BCA_1 or angle BCA is a central angle and the angle BDA_1 or angle BAD is an inscribed angle. Naturly, the two angles have intercepted at the same minor arc $\overset{\frown}{AB}$ on the same circle D. By the Theorem 11-2 and its corollaries, we can construct the following Fig. 11-12 from $m\angle AOB = \theta$ to $m\angle ADB = \frac{1}{4}\theta$

$m\angle ADB = \frac{1}{2}m\angle BCA_1$

$\qquad = \frac{1}{2}m\angle BCA$

$\qquad = \frac{1}{2} \cdot \frac{1}{2}m\angle AOB$

$\qquad = \frac{1}{4}\theta$

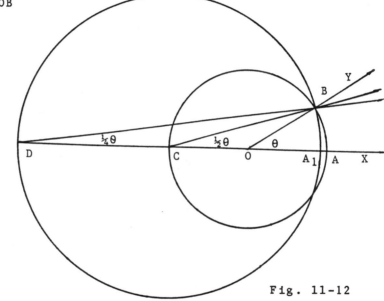

Fig. 11-12

In the same manner, if we repeatly construct one-half angle of the angle BDA_1, the angle AOB will be divided into as small as $(\frac{1}{2})^n m\angle AOB$ (n = 1,2,3,---.) As the result, the following Fig. 11-13 is shown the repeating construction for finding one half angle

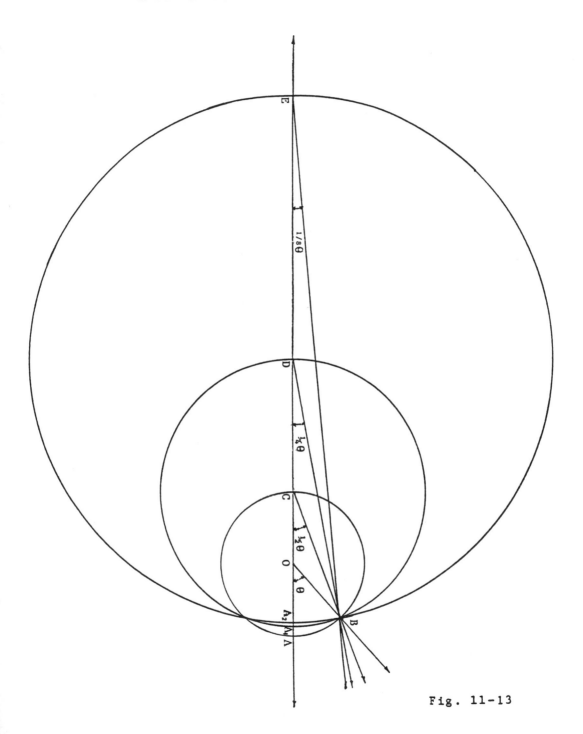

Fig. 11-13

On the Fig. 11-13, we can continuously construct; such as: $\frac{1}{2}m\angle AOB$, $\frac{1}{4}m\angle AOB$, $\frac{1}{8}m\angle AOB$, --- and so on. The point C is the vertex of one-half angle AOB, and the point D is the vertex of one-fourth angle AOB. Between the $\frac{1}{2}m\angle AOB$ and $\frac{1}{4}m\angle AOB$ must exist a vertex of one-third angle AOB in terms of the modern real number theory.[*] For finding and vertifying the vertex of one-third angle AOB, we have made a new postulate in the last section. Each vertex of the bisecting angles must be related to another point on the Euclidean geometric space.

Here, let us investigate the Fig. 11-13 more carefully, all the defined centers of these circles are very closely related to the fixing point B which is passing through by each circle. Actually, in doubling or bisecting an angle, the point B is defined by one side of the given angle and the circle O with a radius OB. However, when we go to double or bisect an angle, the point B can be located easily. Similarly, it is possible for us to base on this principle of doubling and bisecting an angle, and apply to triple and trisect an angle. Curiously, we are going to inquire the relationship between tripling and trisecting an angle in the next chapter.

Evidently, from bisecting or doubling an angle, we have found or discovered a pattern that we employed, a supplementary or vertical angle of the given angle to construct its bisecting or doubling angle indirectly. We do not bisect or double directly to a given angle from its vertex. In the same manner, we are going to focus on tripling or trisecting an angle by using an indirect method to generalize the pattern which we have done and derived in this chapter. Naturally, from the ideas to the pattern, we must produce an example of real mathematical theorems for the entire theoretical discussion.

---

[*] George Cantor: TRANSFINITE NUMBER, Dover Publications, Inc. 1955, pp. 89-91.

CHAPTER III

NEW THEORY OF TRISECTION

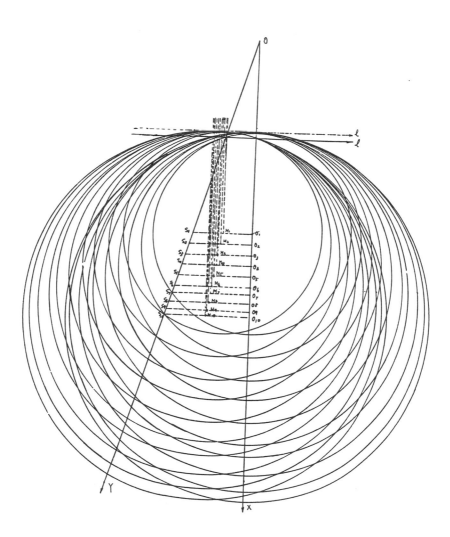

Chapter   III.   New Theorey of Trisection

III-1.   Tripling an Angle

III-2.   Eleven Theorems

III-3.   Trisecting an Angle

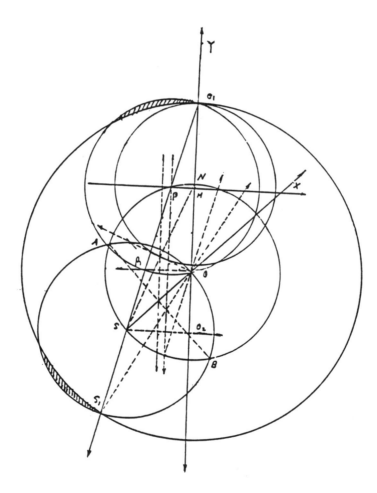

111-1 Tripling An Angle

In the Chapter 11, we have discussed how to double and bisect
an angle by using the relationship between a central angle and
an inscribed angle in the same arc of a circle or a congruent
circle. If a given acute angle is θ ($0 < \theta < \pi/2$), then we can
easily construct its doubling angle 2θ. On the other hand, the
question is: "Can we reverse to construct the original single
angle from the doubled angle?" Here, for answering the question,
it is very important for us to reexamine the doubling and
bisecting an angle from another aspect of the basic relationship
between an exterior angle and its two remote interior angles in
an isosceles triangle. In general property, the measurement of
an exterior angle is equal to the sum measurement of its two
remote interior angles.

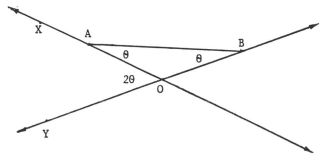

Fig.111-1

In the Fig.111-1 above, if we let a given acute angle XOY be
2θ ($0 < \theta < \pi/2$), then, we can construct an isosceles triangle OAB,
the point A is on the ray $\overrightarrow{OX}$, and the point B is on the ray $\overrightarrow{OY}$
opposite side. Evidently, in the isosceles triangle OAB, since
OA = OB, then, we have; m∠OAB = m∠OBA; but, from an addition,
m∠OAB + m∠OBA = 2θ. Thus, m∠OAB = m∠OBA = θ. As a result,
it is similar to draw a circle with center O, and the circle
intercepts the ray $\overrightarrow{OX}$ at the point A and the opposite ray $\overrightarrow{OY}$ at
the point B. Thus, the triangle OAB is an isosceles triangle.

If we repeatly construct an isosceles triangle BOC with vertex B from the Fig. 111-1, the point C is on the opposite ray $\overrightarrow{OX}$, then, angle BOC is congruent to angle BCO. Clearly, m∠BOC = m∠BCO = 2θ, because the angle BOC is one of the three exterior angles of the isosceles triangle AOB. If we extend $\overrightarrow{CB}$ to be $\overrightarrow{CT}$ (C-B-T), the point T is on the ray $\overrightarrow{CB}$, then, the angle ABT is an exterior angle of the triangle ABC. From the relationship between an exterior angle and its two remote interior angles in the triangle ABC, we have:

$$m∠ABT = m∠BAC + m∠ACB;$$
$$= θ + 2θ$$
$$= 3θ$$

Consequently, the measurement of the angle ABT is to be a tripling angle of the given angle BAC. In other words, the measurement of the angle BAC is one-third of the angle ABT in the following Fig.111-2.

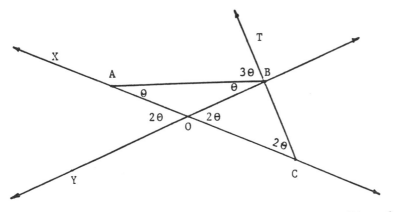

Fig. 111-2

According to the construction in the Fig.111-2, we discover that if we can construct a triangle in which two of the three interior angles have a ratio of the measurement to be 1 to 2, then the measurement of the exterior angle must be a triple angle of its two remote interior angles.

Furthermore, if the 3θ is less than pi (π) radian on the Fig. 111-2, then, we can continue to construct an isosceles triangle CBD as the following Fig.111-3. The point D is on the ray $\overrightarrow{AB}$ (A-B-D) with CB = CD. Also, we extend ray $\overrightarrow{DC}$ to be ray $\overrightarrow{DU}$ (D-C-D), the angle ACU is an exterior angle of the triangle ACD of its two remote angles including angle DAC and angle ADC. As we have:

$$m\angle ACU = m\angle DAC + m\angle ADC$$
$$= \theta + 3\theta$$
$$= 4\theta$$

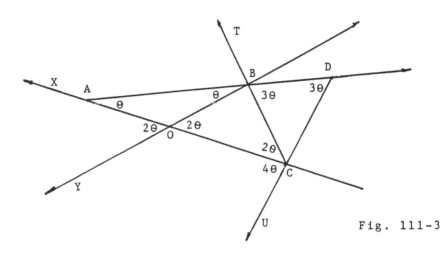

Fig. 111-3

In the same manner, we can construct an angle ACU that is four times measurement of the angle DAC. As the result, the triangle DAC has a ratio of the measurement of two interior angles. If the measurement of 4θ is less than a π radian, then, we can keep to construct 5θ, 6θ, 7θ, ··· nθ (n is a natural number), until the measurement of nθ is equal or less than π radian.

If we let θ radian be the measurement of a very tiny angle, then we continue to construct its double times, triple times, four times, ··· n times (n = 2, 3, 4, ···) drawing. The figure of construction will be a beautiful geometric tree on the Fig.111 4.

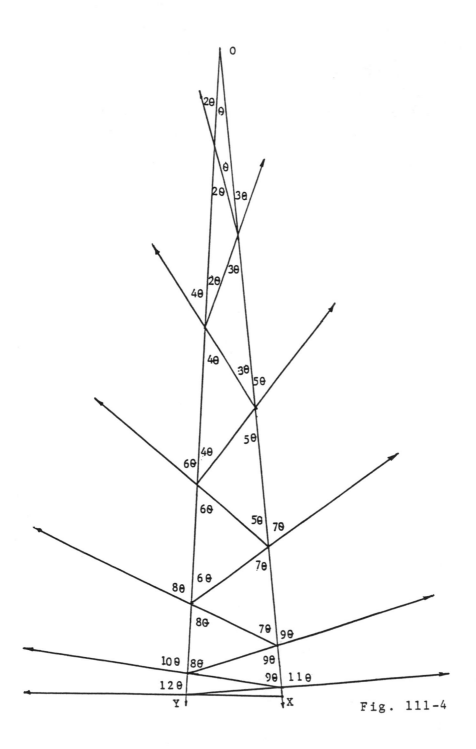

Fig. 111-4

On the Fig.111-4, we have generalized to construct n times angle from a given acute angle θ by repeated constructing an isosceles triangle. Those triangles are not congruent to each other completely, but, they have two congruent sides. If we draw a circle on each vertex, then, all the circles are congruent with equal radii.*

Here, we are going to focus on a new theory of of trisection. It is similar to double and bisect an angle from the begining of this section. If a given acute angle is θ radian. First, we construct an isosceles triangle to double the given acute angle. Then, from the doubling angle 2θ, we repeatly continue to construct an isosceles triangle for tripling angle 3θ as showing Fig.111-2. On the other hand, the question is:"Can we reverse to construct the original single θ from the tripling angle 3θ?" In answering this question, let us look back to the Fig. 111-3 carefully. Originally, the given angle OAB is congruent to the angle OBA because of the isosceles triangle OAB, two base angles have same measurement θ radian. Then, we construct an isosceles triangle OBC with BO = BC, therefore, one exterior angle ABT is to be 3θ. But, from the angle ABT, we can define the point C which is on the opposite ray $\overrightarrow{BT}$, the length of $\overline{BC}$ is an arbitrary segment. Also, we know the segment $\overline{BC}$ is congruent to segment $\overline{BO}$, because the triangle OBC is an isosceles which we have constructed. Now the major problem is how to define the point O. For definding the point O, it is important to apply the new postulate in the previous Chapter 11 for finding a relating point. If we construct a perpendicular line ℓ to the segment $\overline{AB}$ at M from the vertex O of the isosceles triangle AOB, obviously, the point M is the midpoint of the segment $\overline{AB}$. If the point M can be defined, then the point A is the single angle θ which can be defined immediately. Also, the point O can be defined by the collinear with points A and C.

_____
* see the appendix on the page 109.

The point O always is on the line ℓ , the following Fig. 111-5
is deriving form the Fig. 111-2.

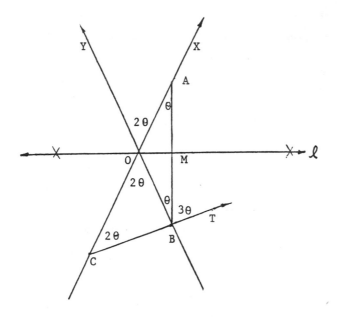

Fig. 111-5

However, the Fig. 111-2 is showing construction from 2θ
backward to θ.   But, in order to generalize tripling angle
from a given an acute angle XOY. If we let O_1 be an arbitrary
point on OX $(O-O_1-X)$ and construct a perpendicular line ℓ to
bisect $\overline{OO_1}$ at point M and meet OY at point P $(O-P-Y)$ respectively.
After that, let PO_1 be a radius and draw a circle with center
O_1 to meet $\overline{OO_1}$ at N, and $\overrightarrow{OY}$ at S.   Obviously, triangle OPO_1 and
and triangle PSO_1 are isosceles triangles.   The measurement
of angle OO_1 T is equal to three times measurement of the given
angle XOY.   All properties of the tripling an angle will
be proved in the next section.   The following Fig. 111-6,
Fig. 111-7, and Fig. 111-8 are to be illustrated in general
of tripling an angle.   On the figures, the points; O, O_1, and
M are fixed points, but the point P is always change in term
of the given angle XOY.

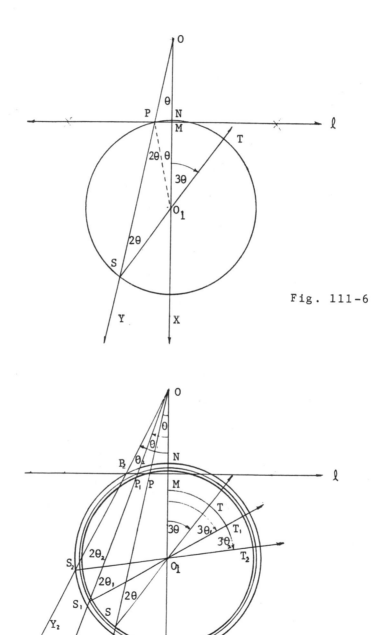

Fig. 111-6

Fig. 111-7

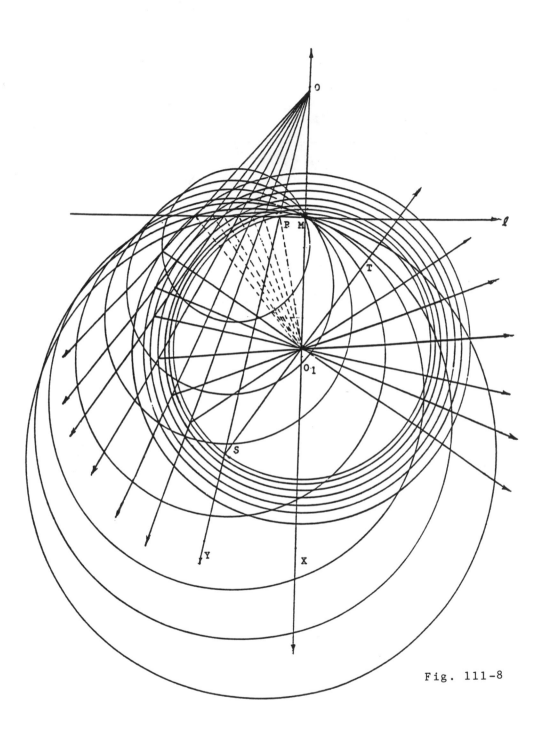

Fig. 111-8

111-2 Eleven Theorems

From the Fig. 111-6, Fig. 111-7, and Fig. 111-8 are
showing the given angle XOY increasing its measurement,
the point P is moving on the line ℓ which is a bisecting
and perpendicular to the segment $\overline{OO_1}$ at the point M. The
three points O, M, and O_1 are fixed without changing locus
on the figute. Since, the construction, there are three
segments are congruent; $PO = PO_1 = SO_1$ thus, the triangle
OPO_1 and triangle PO_1S are isosceles triangles. The ratio
of the measurement of the two interior angles; $m\angle POO_1$ and
$m\angle PSO_1$ is 1 : 2 in the triangle OSO_1. Thus, the measurement
of the exterior angle OO_1T is three times measurement of
the given angle XOY. The truth will be proved in the first
theorem.

Evidently, through eleven theorems for the properties
of tripling an angle, we are searching a point which is
related to the point M. Clearly, on the Fig. 111-8, the
circles are passing through the point M, all the centers
of these circles are the key points.   The question are :
"How do the centers relate to the point M?" and " How can
we state and prove the relationship between the center
of circle and the point M?" From the first Theorem to the
last one will be discussed and serached intensively the
properties between tripling and trisecting an acute angle.
If the key point M can be defined, then, the point O can
be immediately located. because of $MO = MO_1$ At the same time,
the intercepting point P can be situated among the line ℓ .
the ray $\overrightarrow{OS}$, and the circle O_1.

All the following theorems are discovered by seraching
a new approach to the trisection problem. Each theorem will
be discussed, examined, and verified in terms of the axims,
postulates, and properties of the original Euclid's Elements.

NEW THEORY OF TRISECTION

Theorem I.   If an acute angle is defined between 0 and $\pi/3$.
then, the acute angle can be tripled[*].

Construction:

1. Let an acute angle be $\angle XOY$, and the measurement of the acute angle be θ ($0 < 0 < \pi/3$)

2. Pick an arbitrary point 0_1 on the side $\overrightarrow{OX}$.

3. At the midpoint M of the segment $\overline{OO_1}$, construct a perpendicular and bisecting line ℓ , so that it meets side $\overrightarrow{OY}$ at the point P.

4. Connect PO_1.

5. Let point 0_1 be a center and PO_1 be a radius, and draw a circle $\odot O_1$ to meet $\overrightarrow{OY}$ at the point S and $\overline{OO_1}$ at the point N.

6. Draw a ray $\overrightarrow{SO_1}$ to meet the circle O_1 at point T.

7. The angle TO_1O is a tripled angle to the angle XOY.

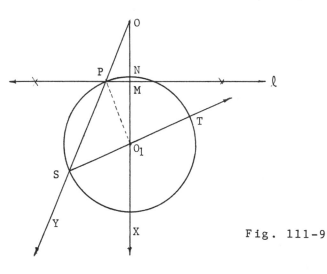

Fig. 111-9

---

* On the Definition 8 and 9 of the Book I of the Elements
8 A plane angle is the inclination to one another of two lines in a plane which meet one another and do not lie in a straight line.
9 And when the lines containing the angle are straight, the angle is called rectlineal (pp.176-78).

Proof By the construction Fig.111-9 of the Theorem, since, the line ℓ is a bisector and perpendicular to the segment $\overline{OO_I}$ at M, also P is a point on the line ℓ and circle O_1. Thus, $\triangle POO_1$ is an issosceles triangle[*]. Hence, we have two base angles which are equal to an isosceles triangle[*]; that is:

$$m\angle POO_1 = m\ PO_1O = \theta$$

Since PO_1 and SO_1 are radii of $\odot O_1$, we have $PO_1 = SO_1$

Obviously, $\triangle PO_1S$ is an isosceles triangle too:
$$m\ O_1PS = m\ O_1SP$$
But, $\angle O_1PS$ is an exterior angle[**] of $\triangle PO_1O$.

Thus, we have:

$$m\angle O_1PS = m\angle O_1SP$$
$$= m\angle POO_1 + m\angle PO_1O$$
$$= \theta + \theta$$
$$= 2\theta$$

And, $\angle OO_1T$ is an exterior angle of $\triangle SO_1O$.

Consequently,

$$m\angle OO_1T = m\angle O_1OS + m\angle O_1SP$$
$$= m\angle POO_1 + m\angle O_1SP$$
$$= \theta + 2\theta$$
$$= 3\theta \qquad\qquad Q.E.D.$$

---

* On the proposition 5 of the Book I of The Elements (p.25). In isosceles triangles the angles at the base are equal to one another, and, if the equal straight lines be produced further the angles under the base will be equal to one another.

** On the Proposition 32 of the Book I of The Elements (pp.316-17). In any triangle, if one of the sides be produced, the exterior angle, is equal to the two interior and opposite angles of the triangle are equal to two right angles.

Theorem II.   By the procedure of constructing a tripled angle
in the Theorem I, the angle OO_1T is tripled by a
given angle XOY as the folowing Fig. 111-10.$\overset{*}{}$
If and only if $m\angle OO_1T = 3m\angle XOY$,
then,

(i) $m\angle O_1NS = m\angle O_1SN = \frac{3}{2} m\angle XOY$;

$m\angle NSP = \frac{1}{2} m\angle XOY$;

$m\angle PSO_1 = m\angle O_1PS = 2 m\angle XOY$;

$m\angle NO_1P = m\angle XOY$;

(ii) P is the intercepting point of $\odot O_1$, line ℓ,
and side $\overrightarrow{OY}$; line ℓ is a perpendicular and
bisector of $\overline{OO_1}$; $\overrightarrow{OY}$ is on side of the given
angle XOY.

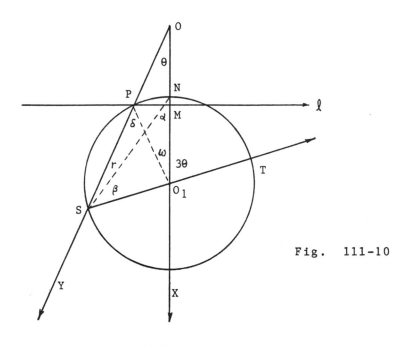

Fig.   111-10

---

* On the Fig.111-10, the point N is intercepted
by the segment $\overline{OO_1}$ and O_1. The point M is on
the ray $\overrightarrow{OX}$ and the midpoint of the segment $\overline{OO_1}$.

Proof: By the Fig. 111-10, we are going to prove the first:

If (i) and (ii) are hold,

then, $m\angle OO_1T = 3m\angle XOY$.

It had been proved in the Theorem I.

Now, we are going to prove the inverse statement:

If $m\angle OO_1T = 3m\angle XOY$, then, (i) and (ii) are hold.

From the given, we let:

$$m\angle XOY = \theta \quad (0 < \theta < \pi/3)$$

Hence, $$m\angle OO_1T = 3\theta;$$

Also, we let:

$$m\angle SNO_1 = \alpha \; ; \qquad m\angle NSO_1 = \beta \; ;$$

$$m\angle NSP = r \; ; \qquad m\angle O_1PS = \delta \; ;$$

$$m\angle NO_1P = \omega \; ;$$

Since $\triangle O_1NS$ is an isosceles triangle,

Thus, we have:

$$\alpha = \beta \text{----------------}(1)$$

But, $\angle OO_1T$ is an exterior angle of the $\triangle SO_1N$

And, we have: $$\alpha + \beta = 3\theta \text{------------}(2)$$

From (1) and (2), we obtain:

$$\alpha = \beta = \frac{3}{2}\theta \text{------------}(3)$$

Or $$m\angle O_1NS = m\angle O_1SN = \frac{3}{2}m\angle XOY$$

Since, $\angle OO_1T$ is also an exterior angle of $\triangle SO_1O$

Thus,

$$(\beta + r) + \theta = 3\theta \quad \text{or} \quad \beta + r = 2\theta \text{------------}(4)$$

From (3) and (4), we have:

$$\frac{3}{2}\theta + r = 2\theta$$

$$r = 2\theta - \frac{3}{2}\theta$$

$$= \frac{4}{2}\theta - \frac{3}{2}\theta$$

$$= \frac{1}{2}\theta \text{------------}(5)$$

That is $m\angle NSP = \frac{1}{2} m\angle XOY$ to be hold.

Since $\angle PSN$ and $\angle PO_1N$ have the same arc $\overarc{PN}$ on the circle $\overset{*}{O}_1$
Thus, we have:

$$\omega = 2\,r$$
$$= 2 \cdot \frac{1}{2}\,\theta$$
$$= \theta \quad \text{------------(6)}$$

Hence, $m\angle NO_1P = m\angle XOY$, and $\triangle PO_1O$ is an isosceles triangle.
The point P is on the line ℓ which is perpendicular and
bisecting to the segment $\overline{OO}_1$.
Also, $\angle O_1PS$ is an exterior angle of $\triangle O_1OP$.
Hence,

$$\delta = \theta + \omega$$
$$= \theta + \theta$$
$$= 2\theta \quad \text{------------(7)}$$

From (7) and (4), we obtain:

$$\delta = 2\theta = \beta + r\,;$$

Obviously, $\triangle PO_1S$ is an isosceles triangle, thus, we have:

$$O_1P = O_1S$$

Consequently, P must be a point on the circle O_1 and $\overrightarrow{OY}$.

Q.E.D.

---

* On the proposition 20 and 21 of the
Book III of The Elements (pp. 46-50).
Proposition 20: In a circle the angle at
the centre is double of the angle at the
circumference, when the angles have the
same circumference as base.
Proposition 21: In a circle the angles in
the same segment are equal to one another.
According to the Proposition 21, if any
number of triangles be constructed on the
same base and on the same side of it, with
equal vertical angles, the vertices will
all lie on the circumference of a segment
of a circle.

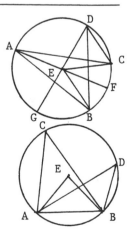

Theorem III.   By the procedure of constructing a triple angle
in the Theorem I, the angle TO_1O is tripled by
a given angle XOY as the following Fig. 111-11
On the figure, the ray $\overrightarrow{ON_1}$ is perpendicular to
$\overline{SN}$ at the point M_1 and meeting the ray $\overrightarrow{OY}$ at N_1.
If $m\angle TO_1O = 3\ m\angle XOY$, then $NO = NN_1 = S\overset{*}{N}_1$

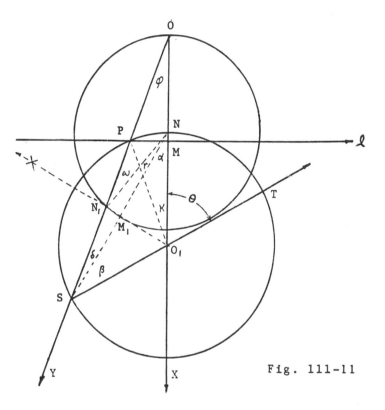

Fig. 111-11

---

* The Theorem III, was discovered before Christmas in 1990.
After Professor T.A. Romberg at National Center for research
in Mathematical Sciences Education of the University of
Wisconsin-Madison, helped me examine and cite my first draft,
he found the original proposition was disproved. Therefore,
the Theorem was restated and reverified, it was an important
turning point for me to keep going on the trisection problem.
In order to appreciate and honor T.A. Romberg. The Theorem III
will be named Romberg-Chen Theorem.

Proof: By the Fig.111-11, first, we connect segments $\overline{PO_1}$, $\overline{NN_1}$, and $\overline{NS}$, then, we let:

$$m\angle O_1NS = \alpha \; ; \qquad m\angle O_1SN = \beta \; ;$$
$$m\angle N_1NS = \gamma \; ; \qquad m\angle N_1SN = \delta \; ;$$
$$m\angle O_1ON = \varphi \; ; \qquad m\angle NN_1O = \omega \; ;$$
$$m\angle NO_1P = \kappa \; ;$$

Now, we are going to prove:

If $\varphi = \frac{1}{3}\theta$, then $NO = NN_1 = SN_1$

By the Theorem II, O_1NS is an isosceles triangle. Thus, we have two base angles are equal;

$$\alpha = \beta = \frac{1}{2}\theta \; \text{---------------}(1)$$

From the given, $\overline{O_1N} \perp \overline{SN}$; M_1 is the midpoint of $\overline{SN}$. By the properties of congruence in two triangles, triangle SM_1N_1 is congruent to the triangle NM_1N_1 Thus, we have:

$$NN_1 = SN_1$$

And,
$$\gamma = \delta \; \text{----------------}(2)$$

Since, $\angle N_1NO_1$ is an exterior angle of $\triangle NON_1$;

Hence,
$$\alpha + \gamma = \varphi + \omega \; \text{----------------}(3)$$

From the given, $\varphi = \frac{1}{3}\theta$, (1) and (2) substitute into the (3), then we have:

$$\frac{1}{2}\theta + \gamma = \frac{1}{3}\theta + \omega \; ;$$
$$\frac{1}{2}\theta - \frac{1}{3}\theta = \omega - \gamma$$
$$\frac{1}{6}\theta = \omega - \gamma \; \text{----------}(4)$$

From (2), NSN_1 is an isosceles triangle, therefore, we can find the relationship among the angles ω , γ , and δ . Continuously, we must prove that triangle NON_1 is an isosceles triangle in terms of two equal base angles.

Here, ω is an exterior angle of the $\triangle NN_1S$.
Thus, by the relationship between exterior angle and interior angle, we have:

$$\omega = \delta + r$$

From (2),
$$\omega = r + r$$

Or,
$$\omega = 2r$$

And,
$$\omega - 2r = \theta \text{ ----------------(5)}$$

From (4), let $r = \omega - \frac{1}{6}\theta$ substitute into (5)
Then, we obtain:

$$\omega - 2(\omega - \frac{1}{6}\theta) = 0$$

$$\omega - 2\omega + \frac{1}{3}\theta = 0$$

$$-\omega + \frac{1}{3}\theta = 0$$

$$-\omega = -\frac{1}{3}\theta$$

$$\omega = \frac{1}{3}\theta \text{ --------------(6)}$$

By (1), (2), (3), (4), (5), and (6), as the result, we obtain:

$$\omega = \varphi = \frac{1}{3}\theta$$

Finally, the triangle N_1NO is an isosceles triangle.[*]

$$NO = NN_1 = SN_1$$

Q.E.D.

---

[*] The triangle N_1NO is an isosceles triangle, the point N_1 always on the ray $\overrightarrow{OY}$ and $\overrightarrow{ON}_1$. The $\overrightarrow{ON}_1$ is always perpendicular to the chord $\overline{SN}$ of circle O_1. If we consider the right triangle NM_1N, when the the given angle XOY is very small, the point N_1 is moving toward M_1. If the angle XOY is increasing its measurement, the point M_1 is moving toward the point N_1.

Theorem IV.   By the procedure of constructing a triple angle
in the Theorem I, the angle TO_1O is tripled by
a given angle XOY as the following Fig. 111-12.[*]
on the figure, P_1 is a intercepting point of $\overrightarrow{OY}$
and $\overrightarrow{O_1P_1}$. The ray $\overrightarrow{O_1P_1}$ is perpendicular to the
ray $\overrightarrow{OX}$; and we let:

$a = OM = O_1M = \frac{1}{2} OO_1 \ ; \quad P = PM \ ;$

$r = OP = O_1P = O_1S = NO_1 \ ;$

If and only if $m \angle TO_1O = 3 \ m \angle XOY$,

then, (i) $MN = r - a$ or $P = \sqrt{(r - a)(r + a)}$

(ii) $2 \ MP = O_1P_1$ or $MP = \frac{1}{2} O_1P_1$

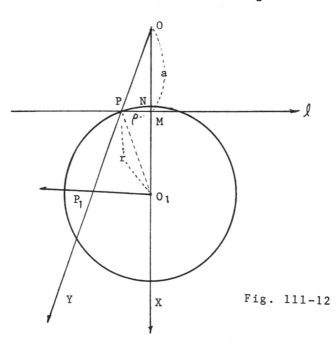

Fig. 111-12

---

* On the Fig.111-12, M is the midpoint of $\overline{OO_1}$
N is intercepted by the ray $\overrightarrow{OX}$ and circle O_1.
The line ℓ is perpendicular and bisecting to
the segment $\overline{OO_1}$.

Proof: By the given and the Fig.111-12 we constructed.
First, we prove:

$$\text{If } m\angle TO_1O = 3m\angle XOY,$$

$$\text{then, (i) and (ii) are hold.}$$

Since, $\ell \perp \overrightarrow{OX}$;

Thus, $\triangle OMP$ and $\triangle O_1MP$ are right triangles.
According to the Pythagorean theorem;
we have:

$$PM^2 = OP^2 - OM^2$$

$$PM^2 = O_1P^2 - O_1M^2$$

Or

$$\rho^2 = r^2 - a^2$$

$$= (r - a)(r + a)$$

$$\rho = \sqrt{(r - a)(r + a)}$$

Hence, ρ is a geometrical mean* of $(r - a)$ and $(r + a)$
Obviously, the (i) is hold.
Next, we are going to prove (ii) is hold.
Since, $\overrightarrow{O_1P_1} \perp \overrightarrow{OX}$; thus, $\triangle OO_1P_1$ is a right triangle.
From the Theorem I, the ray $\overrightarrow{MP}$ is perpendicular to the
ray $\overrightarrow{OX}$. By the parallel Theorem, we have:

$$\overrightarrow{MP} /\!/ \overrightarrow{O_1P_1}$$

And, $\triangle OMP$ and $\triangle OO_1P$ are similar each other.
By the properties of two similar triangles and the
given, we obtain:

$$OM = O_1M = \frac{1}{2} OO_1$$

And,

$$\frac{MP}{OP} = \frac{OM}{OO_1}$$

$$= \frac{OM}{2OM}$$

$$= \frac{1}{2}$$

Or,

$$2MP = O_1P_1$$

$$MP = \frac{1}{2}O_1P_1$$

As the result, (ii) is hold too.

---

* see next page

Inversly, we are going to prove:

$$\text{If} \quad r^2 = \rho^2 + a^2 \quad \text{and} \quad PM = \tfrac{1}{2} O_1P_1 \ ;$$
$$\text{then,} \quad m\angle TO_1O = 3m\angle XOY.$$

From, $O_1P = OP = r$; $O_1M = OM = a$; and $r^2 = \rho^2 + a^2$;

Also, $\triangle O_1MP$ and $\triangle OMP$ are congruent.

Thus, we have:

$$m\angle MOP = m\angle MO_1P$$

By the given, $\qquad\qquad MP = \tfrac{1}{2} O_1P_1$

And, from the Fig. 111-12, circle P and circle O_1 are congruent circles, also, angle O_1PP_1 is one the exterior angle of the triangle O_1PO.

Thus we have: $\qquad m\angle O_1PP_1 = 2m\angle PO_1N$

Or $\qquad\qquad\qquad m \overset{\frown}{NP} = \tfrac{1}{2}m \overset{\frown}{O_1P_1}$

And, $\qquad\qquad\qquad m\angle O_1PP_1 = 2m\angle NOP$

$$= 2m\angle MOP$$

$$= 2m\angle XOY$$

The $\triangle O_1PS$ is an isosceles triangle, thus two base angles are equal to each other.

$$m\angle O_1PP_1 = m\angle O_1SP$$

But, $\angle TO_1O$ is an exterior angle of the $\triangle O_1OS$;

Consequently, $\qquad m\angle TO_1O = m\angle XOY + m\angle PSO_1$;

$$= m\angle XOY + 2m\angle XOY;$$

$$= 3m\angle XOY.$$

Q.E.D.

---

* Geometric Mean is to construct a segment whose length is the mean proportion between the lengths of two given segments. If given two segments are a and b, then a segment of length x such that $a:x = x:b$ or $x^2 = ab$. The x is called the geometric mean of a and b.

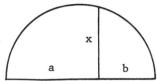

Theorem V.   By the procedure of constructing a triple angle
in the Theorem I, the angle TO_1O is tripled by
a given angle XOY as the following Fig. 111-13.
P is the intercepting point of ℓ , $\overrightarrow{OY}$, and $\odot O_1$;
ℓ is a bisector of $\overline{OO}_1$ and $\ell \perp \overline{OO}_1$:
M is the midpoint of $\overline{OO}_1$;
P_1 is the intercepting point of $\overrightarrow{O_1P_1}$ and $\overrightarrow{OY}$;
M_1 is the midpoint of $\overline{O_1P_1}$;
$m\angle TO_1O = 3m\angle XOY$;
If segment $\overline{PM}_1$ is extended to meet $\odot O_1$ at P_2,
then, $PP_2 = OO_1$ and $m\angle O_1PP_2 = m\angle O_1P_2P = m\angle XOY$.

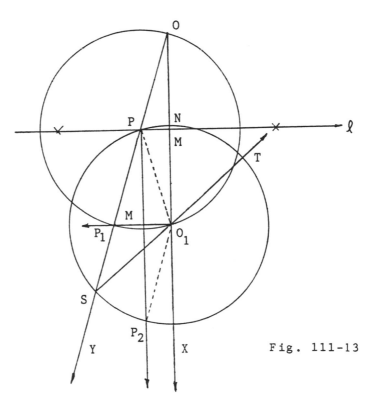

Fig. 111-13

Proof: Connect the segments $\overset{\frown}{PO_1}$ and $\overline{O_1P_2}$;
By the given and the Theorem IV;
We have: $PM = \frac{1}{2} O_1P_1$;
Since M_1 is the midpoint of the segment $\overline{O_1P_1}$;
Thus, $PM = O_1M_1$ -------------(1)
By the construction a triple angle in the Theorem I;
$$\ell \perp \overline{OO_1}; \overline{O_1P_1} \perp \overline{OO_1};$$

Thus, we have:
$$m\angle O_1MP = m\angle MO_1P = \frac{1}{2} \text{-------(2)}$$
And, by the Theorem IV;
$$\overline{PM} /\!/ \overline{O_1M_1} \text{-------------(3)}$$
From (1), (2), and (3), the quadrilateral MPM_1O_1 is
a rectangular. Hence, we have:
$$\overline{PM_1} /\!/ \overline{MO_1}$$
$$MO_1 = PM_1 \text{--------------(4)}$$
By the given, two circles $\odot P$ and $\odot O$ are congruent
each other, and P is the center of the circle P, O_1
is the center of the circle O_1.
Thus, the radii are equal in a circle or the equal or
congruent circles.

$$MO_1 = MO \quad ; \quad PM_1 = P_2M_1$$

$$2MO_1 = OO_1 \quad ; \quad 2PM_1 = P_2P_1 \text{-----(5)}$$

From (4) and (5), we obtain:

$$2 \cdot MO_1 = 2 \cdot PM_1$$

$$OO_1 = PP_2$$

And, two triangles; $\triangle OPO_1$ and $\triangle O_1PP_2$ are isoceles
and congruent triangles. Hence, the base angles are
congruent each other.
$$m\angle O_1PP_2 = m\angle O_1P_2P = m\angle PO_1O = m\angle XOY$$
$$\text{Q.E.D.}$$

Theorem VI.   By the procedure of constructing a triple angle
in the Theorem I, the angle TO_1O is tripled by
a given angle XOY as the following Fig. 111-14.
The circle P is with center at P and radius $\overline{PO_1}$;
The circle S is with center at S and radius $\overline{SO_1}$;
$\odot$ P and $\odot$ S are intercepting at the two points
O_1 and R_1; $m\angle TO_1O = 3m\angle XOY$;
If the circle S meets the side $\overrightarrow{OY}$ of $\angle XOY$ at S_1,
then, $OO_1 = O_1S_1$ and $m\angle SS_1O_1 = m\angle POO_1 = \frac{1}{3}m\angle XOY$

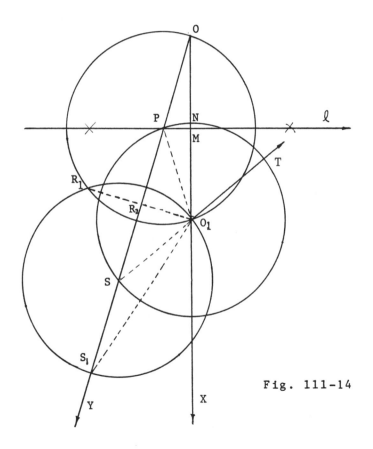

Fig. 111-14

Proof: By the constructing of the Theorem I and the given; we connect segments $\overline{SO_1}$, $\overline{PO_1}$, and $\overline{O_1R_1}$.

Also, we name R_2 is the intercepting point of $\overline{O_1R_1}$ and $\overrightarrow{OY}$. Since PO_1 is the radius of the circle P and SO_1 is the radius of the circle S. Thus, $\odot P$, $\odot S$, and $\odot O_1$ are equal circles, and all the length of radii of the circles are equal, that are:

$$PO_1 = SO_1 \; ; \; PR_2 = SR_2 \; ; \; PO_1 = SS_1 \quad ----(1)$$

Since, $\odot P$ and $\odot S$ are intercepting at the points O_1 and R_1, thus, we have:

$$m\angle PR_2O_1 = m\angle SR_2O_1 = \frac{1}{2}\pi \quad ------------(2)$$
$$PR_2 = SR_2 \quad ------------------------(3)$$

From (1) and (3), we have:

$$R_2P + PO = R_2S + SS_1$$

Or, $\qquad R_2P = R_2S \quad ------------------------(4)$

Also, $\overline{O_1R_2}$ is an common side of the two adhesive right triangles $\triangle O_1R_2O$ and $\triangle O_1R_2S$, thus, we have:

$$O_1R_2 = O_1R_2 \quad ----------------------(5)$$

From (2), (4), and (5), we obtain that $\triangle O_1R_2O$ and $\triangle O_1R_2S$ are congruence.** Hence, the corresponding sides and angles are congruent each other, that are:

$$OO_1 = S_1O_1 \; , \; m\angle SS_1O_1 = m\angle POO_1 = \frac{1}{3} m\angle XOY.$$

$$\text{Q.E.D.}$$

---

* If two equal circles $\odot P$ and $\odot S$ meet two points O and R, the quadrilateral OPRS is a rhombus. One of the property of rhombus is the two diagonals are perpendicular and bisect each other.

** On the proposition 4 of the Book I of The Elements (p.247) if two triangles have the two sides equal to two sides respectively, and have the angles contained by the equal straight lines equal, they will also have the base equal to the base, the triangle will be equal to the triangle, and the remaining angles wil be equal to the remaining angles respectively, namely those which the equal sides subtend.

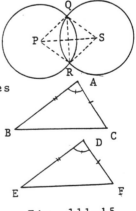

Fig. 111-15

Theorem VII. By the procedure of constructing a triple angle
in the Theorem I, the angle TO_1O is tripled by
a given angle XOY as the following Fig. 111-16
On the figure, three circles $\odot$ P, $\odot$ O_1, and $\odot$ S
are congruent each other. And there is:

   $m\angle TO_1O = 3m\angle XOY$.

If $\odot$ P meets $\odot$ O_1 at the points A and B;

   $\odot$ P meets $\odot$ S at the points O_1 and C;

   $\odot$ O_1 meets $\odot$ S at the points D and E;

then, (i) $\overleftrightarrow{AB}$, $\overleftrightarrow{CO_1}$ and $\overleftrightarrow{DE}$ meet at a point G;

    (ii) P, O_1, and S are lie on the $\odot$ G.

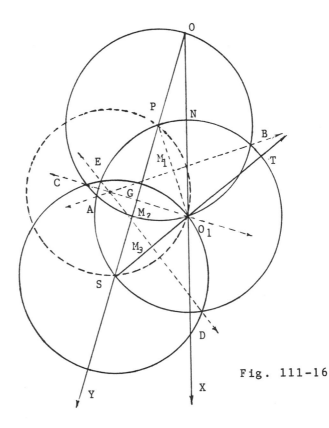

Fig. 111-16

Proof: By the construction of the Theorem I and the given;
we name:

M_1 is the intercepting point of $\overleftrightarrow{AB}$ and $\overline{PO_1}$;

M_2 is the intercepting point of $\overleftrightarrow{CO_1}$ and $\overline{PS}$;

M_3 is the intercepting point of $\overleftrightarrow{DE}$ and $\overline{SO_1}$;

Since, $\odot P$ meets $\odot O_1$ at the point A and B, also by
the Theorem VI, we have:

$$\overleftrightarrow{AB} \perp \overline{PO_1}$$
$$O_1M_1 = PM_1$$
$$AM_1 = BM_1 \ \text{----- -- -----(1)}$$

Since, $\odot P$ meets $\odot S$ at the points O_1 and E.

And, $\odot O_1$ meets $\odot S$ at the points D and E.

At the same reason above, we have:

$$\overleftrightarrow{CO_1} \perp \overline{PS}$$
$$O_1M_2 = CM_2$$
$$PM_2 = SM_2 \ \text{--- ----------(2)}$$

And,
$$\overline{DE} \perp \overline{SO_1}$$
$$O_1M_3 = SM_3$$
$$EM_3 = DM_3 \text{-------------(3)}$$

From (1), (2), and (3), we obtain:

M_1 is the midpoint of $\overline{O_1P}$ and $\overleftrightarrow{AB} \perp \overline{PO_1}$;

M_2 is the midpoint of $\overline{PS}$ and $\overleftrightarrow{CO_1} \perp \overline{PS}$;

M_3 is the midpoint of $\overline{SO_1}$ and $\overleftrightarrow{ED} \perp \overline{SO_1}$;

As the result, $\overline{AB}$, $\overline{CO_1}$ and $\overline{DE}$ are bisectors and
perpendicular to three sides of the triangle PSO_1
respectively. Hence, the three lines $\overline{AB}$, $\overleftrightarrow{CO_1}$, and
$\overleftrightarrow{DE}$ are concurrent at the point G. The point G is
called the incenter because it is the center of a
a circle inscribed in the triangle PSO_1, in the
other words, the perpendicular bisectors of three
sides of a triangle intersect in one point (called
the circumcenter), which is equidistant from the
three vertices.   Q.E.D.

Theorem VIII.   By the procedure of constructing a triple angle
in the Theorem I, the angle TO_1O is tripled by
a given angle XOY as the following Fig. 111-17.
$\odot O$, $\odot P$, and $\odot O_1$ are congruent circles. The $\odot O$
meets line ℓ at the point N_1.   And we draw a
parallel ray to ray $\overrightarrow{OX}$ from the point S, the
ray meets B with the line ℓ and S_1 with $\odot O_1$.
If we extend the ray $\overrightarrow{SO_1}$ or ST to meet the line
ℓ at T_1. Also, we connect $\overline{OT}_1$, $\overline{OS}_1$, $\overline{S_1N_1}$, and $\overline{SN}$.
Then (i) $\triangle ST_1S_1$ is an isosceles triangle, three
points T_1, O, and S_1 are colinear.
(ii)   The point I must be a center of the
inscribed circle of the $\triangle ST_1S_1$.

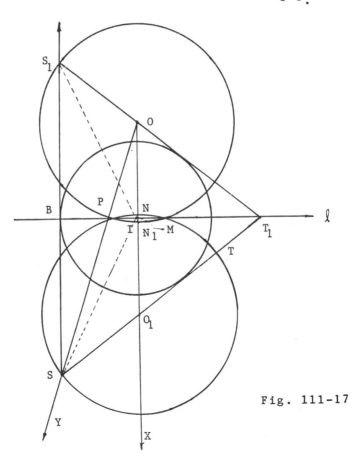

Fig. 111-17

Proof: By the constructing of the Theorem I and the given condition, first, let us prove (i).

Since, line ℓ is perpendicular and bisect to $\overline{OO_1}$. Thus, triangle T_1OO_1 is an isosceles triangle. Also, we have:

$$m\angle OMT_1 = m\angle O_1MT_1 = \frac{1}{2}\pi$$

$$m\angle MOT_1 = m\angle MO_1T_1 \quad \text{------------(1)}$$

$$OT_1 = O_1T_1 \quad \text{--------------(2)}$$

Since, $\overline{SS_1} /\!/ \overline{OO_1}$ and $\overrightarrow{SS_1}$ intercept the line ℓ at the point B. Thus, the corresponding angles are equal:

$$m\angle S_1BT_1 = m\angle SBT_1 = \frac{1}{2}\pi$$

$$m\angle MO_1T_1 = m\angle BST_1 \quad \text{--------------(3)}$$

$$\frac{T_1O}{OS_1} = \frac{T_1O_1}{O_1S} \quad \text{--------------(4)}$$

Or

$$\frac{T_1O + OS_1}{OS_1} = \frac{T_1O_1 + O_1S_1}{O_1S} \quad \text{---------(5)}$$

From (2) $\quad OT_1 = T_1O_1$ substitute into (4)

Hence, we have $\quad OS_1 = O_1S$ and substitute into (5), hence, $T_1O + OS_1 = T_1O_1 + O_1S$

Or $\qquad\qquad T_1S_1 = T_1S$

As the result, the point S_1, O, and T_1 must be colinear, and the triangle ST_1S_1 is an isosceles triangle.

Second, let us prove (ii);

By the construction and given, $\overline{ON_1}$ and $\overline{SO_1}$ are radii of the $\bigodot O_1$. Thus, we have:

$$NO_1 = SO_1 \quad \text{------------------(6)}$$

And the triangle NSO_1 is an isosceles triangle;

$$m\angle SNO_1 = m\angle NSO_1 \text{--------------(7)}$$

But, the angle OO_1T is an exterior angle of NSO_1.

Hence, $\qquad m\angle SNO_1 + m\angle NSO_1 = m\angle OO_1T_1$;

From (7), $\qquad\qquad 2m\angle NSO_1 = m\angle OO_1T_1$;

Or, $\qquad\qquad\qquad m\angle NSO_1 = \frac{1}{2}\, m\angle OO_1T_1$ ----------(8)

since, the ray $\overrightarrow{SS_1}$ is parallel to $\overrightarrow{OO_1}$, thus, all the corresponding angles are equal:

$$m\angle OO_1T_1 = m\angle S_1SO_1 \text{ -------------(9)}$$

let (9) substitute into (8), therefore, we obtain:

$$m\angle NSO_1 = \frac{1}{2}\, m\angle S_1SO_1$$

Obviously, segment $\overline{SN}$ is a bisector of the angle S_1SO_1.
At the same manner, we also obtain; segment $\overline{NS_1}$ is a
bisector of the angle SS_1O. Also, from the given, the
line ℓ is a bisector of the angle ST_1S_1. By the basic
definition of the center of an inscribed circle of an
arbitrary triangle. The segments $\overline{SN}$, $\overline{SN_1}$ and the line ℓ
are to meet at the point I must be the center* of inscribed
circle of the $\triangle ST_1S_1$. $\qquad\qquad$ Q.E.D.

---

* On the proposition 4 of the BOOK IV of The Eléments (p.·85)
that is: In a given triangle to inscribed a circle. Let ABC
be the given triangle; thus it is required to inscribe a
circle in the triangle ABC. And we
let the angle ABC, ACB be bisected by
the straight lines BD, CD and let these
meet one another at the point D; from
D let DE, DF, DG be drawn perpendicular
to the straight lines AB, BC, CA. Now,
since the angle ABD is equal to the
angle CBD, and the right angle BED is
also equal to the right angle BFD.$\triangle$EBD,
$\triangle$FBD are two triangle having two triangles equal to two
angles and sides equal to one side namely that subtending
one of the equal angles, which is BD common to the triangles;
therefore they will also have the remaining sides equal to
the remaining sides; therefore DE is equal to DF. For the
same reason, DG is also equal to DF. As the result, the
circle inscribed the trianlge ABC with center D.

Theorem IX.   By the procedure of constructing a triple angle
in the Theorem I, the angle TO_1O is tripled by a
given angle XOY as the following Fig. 111-18.
S_2 is the intercepting point of $\overleftrightarrow{SS_2}$ and $\odot S$;
S_3 is the intercepting point of $\overrightarrow{SS_3}$ and $\odot O_1$;
P_1 is the intercepting point of $\overrightarrow{OY}$ and $\odot S$;
P_2 is the intercepting point of $\overrightarrow{NS}$ and $\odot S$;
P is the intercepting point of line ℓ and $\odot O_1$;
Also, $\ell \perp OX$, $\overleftrightarrow{SS_3} /\!/ \overrightarrow{OX}$, and $\overleftrightarrow{SS_2} /\!/ \overrightarrow{OX}$.
If and only if $m\angle XOY = \frac{1}{3}\ m\angle TO_1O$,
then, $m\overparen{PS_3} = 2m\overparen{PN}$ or $m\overparen{P_1S_2} = 2m\overparen{P_1P_2}$

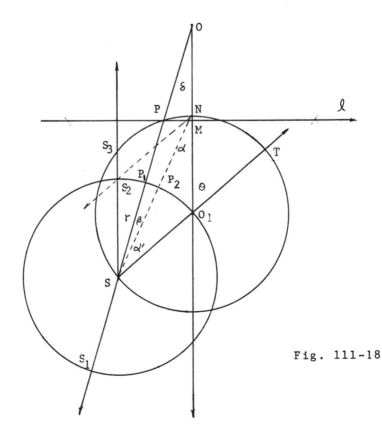

Fig. 111-18

Proof: By the given on the figure, first, we prove:

If $m\angle XOY = \frac{1}{3} m\angle TO_1O$,

then, $m \overarc{PS_3} = 2m\overarc{PN}$ or $m\overarc{P_1S_2} = 2m\overarc{P_1P_2}$

On the circle O_1, since $\overleftrightarrow{SS_3} // \overrightarrow{OX}$;

Thus, $\qquad\qquad m\angle PSS_3 = m\angle XOY$ -----------(1)

By the Theorem II, we have:

$$m\angle NSP = \frac{1}{2} m\angle XOY$$

Or $\qquad\qquad 2m\angle NSP = m\angle XOY$ -----------(2)

From (1) and (2);

$$m\angle PSS_3 = 2m\angle NSP$$

By the definition, the angles are measured by the corresponding arcs, thus, we have:

$$\frac{1}{2} m\overarc{PS_3} = 2 \cdot \frac{1}{2} m\overarc{PN}$$

Or, $\qquad\qquad m\overarc{PS_3} = 2m\overarc{PN}$

In the same manner, on the circle S, we obtain:

$$\overarc{PS_2} = 2m\overarc{P_1P_2}.$$

Second, we reversely prove:

If $m\overarc{PS3} = 2m\overarc{PN}$ or $m\overarc{PS_2} = 2m\overarc{P_1P_2}$,

then, $\qquad\qquad m\angle XOY = \frac{1}{3} m\angle TO_1O$

We let: $\qquad m\angle TO_1O = \theta$;

$\qquad\qquad m\angle SNO_1 = \alpha$; $m\angle NSO_1 = \alpha'$;

$\qquad\qquad m\angle P_2SP_1 = \beta$; $m\angle NSP = \beta'$;

$\qquad\qquad m\angle P_1SS_3 = \tau$; $m\angle XOY = \delta$;

Since, $\angle TO_1O$ is an exterior angle of the isosceles triangle SNO_1, thus, we have:

$$\alpha = \alpha' = \frac{1}{2} \theta \text{ ---------(3)}$$

But, the angle α is an exterir angle of the $\triangle NSO$.

Hence, we obtain:

$$\delta + \beta = \alpha$$

By arc of measuring angle, we have:

$$m\overset{\frown}{PS}_3 = 2m\overset{\frown}{PN} \quad \text{or} \quad m\overset{\frown}{P_1S_2} = 2m\overset{\frown}{P_1P_2};$$

Or

$$\beta = \frac{1}{2}\gamma \quad \text{------------}(4)$$

And, from the given;

$$\overleftrightarrow{SS_3} /\!/ \overrightarrow{OX}$$

Thus, two alternate interior angles are equal to one another;

$$\gamma = \delta \text{-----------------}(5)$$

But, the angle θ is an exterior angle of $\triangle SOO_1$

Hence, $\qquad\qquad \delta + \beta + \alpha' = \theta$

From (3), (4), and (5), we obtain:

$$\delta + \beta + \alpha' = \theta$$

$$\delta + \frac{1}{2}\gamma + \frac{1}{2}\theta = \theta$$

$$\delta + \frac{1}{2}\delta + \frac{1}{2}\theta = \theta$$

$$\frac{3}{2}\delta = \frac{1}{2}\theta$$

$$3\delta = \theta$$

$$\delta = \frac{1}{3}\theta$$

Finally, we obtain; $\qquad m\angle XOY = \frac{1}{3}m\angle TO_1O$

$$\text{Q.E.D.}$$

---

* See the Theorem II-1 & 11-2, the two theorems can be derived three useful corollaries for drawing geometric construction.

Corollary 1. An angle inscribed in a semicircle is a right angle.

Corollary 2. If a quadrilateral is inscribed in a circle then two pair opposite angles are formed by a supplementary.

Corollary 3. If two inscribed angles intercept the same arc or equal arcs in the same circle or equal circles, then the angles are equal.

Theorem X. By the procedure of constructing a triple angle
in the Theorem I, the angle TO_1O is tripled by
a given angle XOY as the following Fig. 111-19.
$\odot O_1$ and $\odot S$ are intercepting at the points A, B;
Segments $\overline{AB}$ and $\overline{SO_1}$ meet at the point C;
Ray $\overrightarrow{SO_2}$ is perpendicular to ray $\overrightarrow{OX}$ at O_2;
If we construct a parallel line to $\overrightarrow{OX}$ from the
point C, the parallel line meets $\overrightarrow{SO_2}$ and $\overrightarrow{SO}$ at
the points M_1 and M_2 respectively.
Then (i) The point M_1 is the midpoint of $\overline{SO_2}$.
            The point M_2 is the midpoint of $\overline{SO}$.
        (ii) $CM_1 = \frac{1}{2} O_1O_2$ and $CM_2 = \frac{1}{2} OO_1$

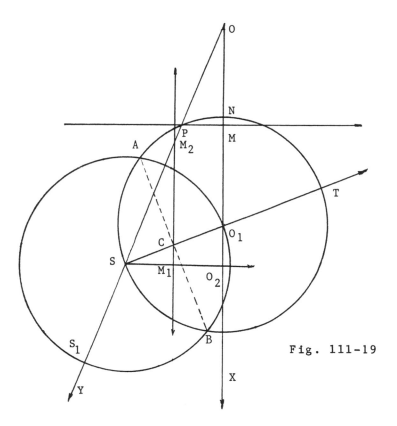

Fig. 111-19

Proof: By the given and the Theorem VI, we are going to prove that (i) and (ii) are hold.

Since, two circles $\odot O$ and $\odot S$ are equal circles.

Thus, two segments $\overline{AB}$ and $\overline{SO_1}$ are perpendicular and bisecting each other at the point C.

Hence, we have:

$$SC = CO_1 \text{ or } SC = \frac{1}{2} SO_1 \quad \text{--------(1)}$$

And, $\qquad\qquad AC = BC$

By the constructed figure,

since, the line $\overleftrightarrow{CM_1}$ or $\overleftrightarrow{CM_2}$ are parallel to $\overrightarrow{OX}$.

Therefore, in the corresponding sides of the two triangles $\triangle SO_2O_1$ and $\triangle SOO_1$ are formed proportion[*],

we have:

$$\frac{SC}{SM_1} = \frac{CO_1}{M_1O_2} \text{ and } \frac{SC}{SM_2} = \frac{CO_1}{M_2O} \quad \text{-------(2)}$$

From (1) substitute into (2), consequently,

we have: $\qquad SM_1 = M_1O_2 \text{ and } SM_2 = M_2O \text{ ;}$

So that, $\quad M_1$ is the midpoint of $\overline{SO_2}$ and,

$\qquad\qquad M_2$ is the midpoint of $\overline{SO}$.

At the same reason of the proportion above, hence,

we have: $\qquad \frac{SC}{CM_1} = \frac{SO_1}{O_2O_1} \text{ and } \frac{SC}{CM_2} = \frac{SO_1}{O_1O} \quad \text{------(3)}$

From (1) and (3), we finally obtain :

$$CM_1 = \frac{1}{2} O_2O_1 \text{ and } CM_2 = \frac{1}{2} OO_1$$

$$\text{Q.E.D.}$$

---

[*] On the proposition 2 of the BOOK VI of The Elements (p.194); if a straight line be drawn parallel to one of the sides of a triangle, it will cut the sides of the triangle proportionally; the line joining the points of the section will be parallel to all the remaining sides of the triangle.

Theorem XI. By the procedure of construction a triple angle
in the Theorem I, the angle TO_1O is tripled by
a given angle XOY as the following Fig. 111-20
21, and 22.[*]

If (i) $m\angle XOY = \frac{1}{3} m\angle TO_1O$;

(ii) PM is the geometric mean of $(MO_1 + NO_1)$
and MN. $\overline{PM} \perp \overline{NO_1}$; $NO_1 = r$ $(r>o)$;
$MO_1 = a$ $(o<a<r)$; $PM = \rho$; and $\rho = r^2 - a^2$;
P is the point on the line ℓ , O_1, and
$\overrightarrow{OY}$. $\ell \perp \overline{OO_1}$; M is the midpoint of $\overline{OO_1}$ on
the line .

(iii) M_6 is the midpoint of $\overline{OP_1}$ and $\overline{OP_1} \perp \overline{OO_1}$;
$O_1P_1 = 2 \cdot O_1M_6$; $\overline{PM_6} /\!/ \overrightarrow{OO}_1$; $\overleftrightarrow{M_6P} /\!/ \overrightarrow{OO}_1$;
P' is the intercepting point of $\overleftrightarrow{NP}$ and
$\overleftrightarrow{M_6P_7}$. $NP' = \ell'$; the line ℓ' is through the
points N, P', and P_{11}: $\overleftrightarrow{NP'} /\!/ \ell$;

(iv) S is intercepted by the ray $\overrightarrow{OY}$ and the
opposite ray $\overrightarrow{O_1T}$.

(v) M_4 is the midpoint of $\overline{SO_2}$.
O_2 is the intercepting point of $\overrightarrow{SO_2}$ and
$\overrightarrow{OX}$. $\overline{SO_2} \perp \overrightarrow{OX}$;

(vi) P_7 is the intercepting point of $\overleftrightarrow{M_4P_7}$ and
$\overleftrightarrow{MP}$ on the Fig.111-20.

(vii) P_{11} is the intercepting point of $\overleftrightarrow{M_4P_{11}}$ and
$\overleftrightarrow{NP'}$ on the Fig. 111-21 and 22.

Then, the point M and P_{11} are on the circle M_4, and
the point P_7 and N are on the circle O_2

---

* For the clear vision, Fig. 111-20, 21, & 22, will
be designed on the next three pages. The key point
M is related to the points S, M_6, P_7, O_1, P_1, P' N,
and P_1 in two rectangles. We are going to prove the
Point M and P' are on the M_6 but they are relating
to the given points S and N. This Theorem XI will be
called the last theorem in the entire theory.

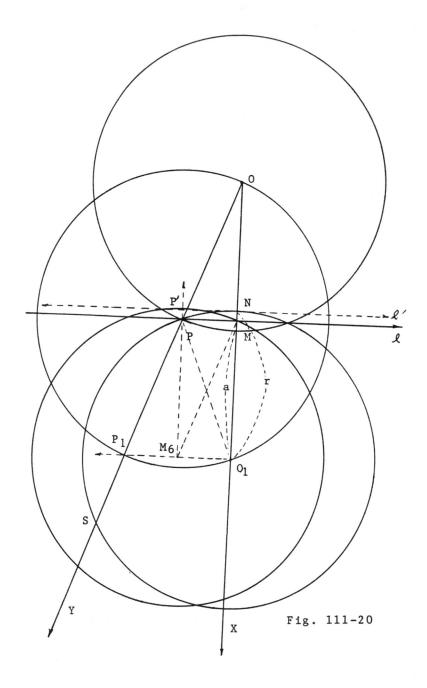

Fig. 111-20

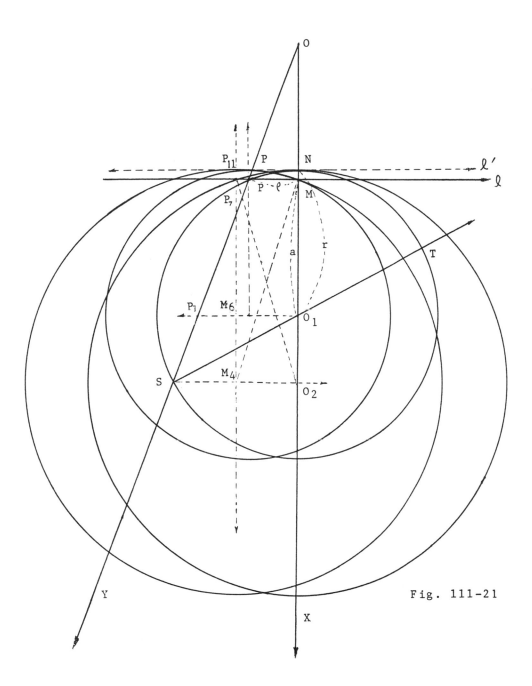

Fig. 111-21

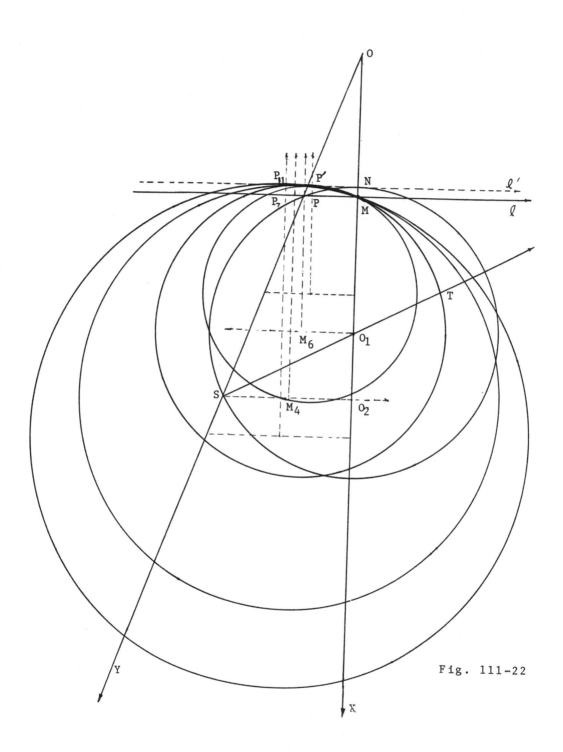

Fig. 111-22

Proof: For the better understanding, the process of proof will be described by three figures including Fig.111-20, 21, First, according to the constructing a triple angle and the given on the Fig.111-20.

Since, polygon $P'M_6O_1N$ and polygon PM_6OM are rectangles. Thus, we have:

$NO_1 = PM_6 = r$; $PM = P'N = M_6O_1 = \rho$; $PM_6 = MO_1 = a$;

Now, we are going to prove:

$$MM_6 = PO_1 = NO_1 = M_6P' = r ;$$

In the two right triangles $\triangle MO_1M_6$ and $\triangle PM_6O_1$, by the given, we have:

$$MO_1 = PM_6 = a ;$$
$$O_1M_6 = O_1M6 = \rho ;$$
$$m\angle MO_1M_6 = m\angle PM_6O_1 = \frac{1}{2}\pi ;$$

Hence, the two right triangles $\triangle MO_1M_6$ and $\triangle PM_6O_1$ are congruent triangles. Therefore, we obtain:

$$MM_6 = PO_1 = NO_1 = M_6P' = r$$

As the result, if we let M_6 be a center, $M_6P' = r$ be a radius, and draw a circle M_6. The point M and P' must be on the circumference of the M_6 because of the same radii. In the other words, we can define the point M from the Points P' and M_6. In the previous theorems, the point M is located exactly at the midpoint of $\overline{OO_1}$. The point M_6 is situated exactly at the midpoint of $\overline{O_1P_1}$. The point P_1^* is the intercepting point of $\overrightarrow{O_1P_1}$ and $\overrightarrow{OY}$.

---

* The point P_1 is unknow point when we reverse to trisect the angle TO_1O. We only know three points O_1, N, and S. We are going to locate the points M, O, and P. The point P is on the side $\overrightarrow{OY}$ offinding angle XOY, also the point S is on the side $\overrightarrow{OY}$ too. Therefore, we are going to continue to prove the point S is the original point to relate to all key points M. O, and P.

Proof: Second, from the Fig.111-20 t0 21 and 22, we draw a
ray $\overrightarrow{SO_2}$ perpendicular to the line $\overline{OO_2}$ and meet at the
point O_2. Also, we let M_4 be the midpoint of the $\overline{SO_2}$.
Next, from the point M_4, we construct a parallel line
with $\overline{NO_2}$ or $\overline{NO_1}$ to meet lines ℓ and ℓ' at the points P_7
and P_1 respectively. Therefore, we obtain two polygons
which are: polygon $NO_2M_4P_{||}$ and polygon MO_1M_6P. Clearly,
both of the polygons are rectangular.

Now, we consider the four rectangles which are: $\square MO_1M_6P$,
$\square NO_1M_6P'$, $\square MO_2M_4P_7$, and $\square NO_2M_4P_{||}$. As we have proved
previously;

$$PO_1 = MM_6 = M_6P' = NO_1 = r \quad -----------(1)$$

And, $$PM_6 = MO_1 = a \quad ------------------(2)$$

In the rectangles $\square MO_2M_4P_7$ and $\square NO_2M_4P_{||}$ by the given
we have:

$$MO_2 = P_7M_4$$
$$M_4O_2 = O_2M_4$$
$$m\angle MO_2M_4 = m\angle P_7M_4O_2 = \frac{1}{2}\pi$$

Thus, two right triangles $\triangle MO_2M_4$ and $\triangle P_7M_4O_2$ are similar
and congruent each other. Hence, we have:

$$MM_4 = P_7O_2 \quad --------------------(3)$$

If we let $O_1O_2 = k$ and from (1) and (2), then, we have:

$$P_7M_4 = MO_2 = MO_1 + O_1O_2 = a + k \quad --------(4)$$

$$P_{||}M_4 = NO_2 = NO_1 + O_1O_2 = r + k \quad --------(5)$$

Since P_7 is on the line which is perpendicular to $\overline{OO_2}$
or $\overline{OO_1}$ or $\overline{O_1O_2}$. Previously, we have proved the Theorem
V, the MP_7 is the geometric mean of $(MO_2 + NO_2)$ and MN.
From (3) we need to prove MM_4 and P_7O_2 are equal to NO_2
or $P_{||}M_4$ in the following page.

Proof: Third, if we let $MP_7 = M_4O_2 = \rho'$, also, by the definition of geometric mean, we have:

$$|MP_7|^2 = (MO_2 + NO_2) \cdot MN$$

or

$$|M_4O_2|^2 = (MO_2 + NO_2) \cdot MN \text{ -----------}(6)$$

From (4) and (5), we have:

$$|P_7M|^2 = |M_4O_2|^2 = \big[(r + a) + (a + k)\big] \cdot \big[(r + k) - (a + k)\big]$$

$$= (r + a + 2k)(r-a)$$

$$= (r + a)(r - a) + 2k(r - a)$$

$$= r^2 - a^2 + 2kr - 2ka \text{ --------------}(7)$$

By the Pythgorean Theorem and the right triangle MO_2M_4, also, from (4), (5), (6) and (7), we have:

$$|M_4M|^2 = MO_2{}^2 + M_4O_2{}^2$$

$$= (a + k)^2 + (r + a + 2k)(r - a)$$

$$= a^2 + 2ka + k^2 + r^2 - a^2 + 2kr - 2ka$$

$$= r^2 + 2kr + k^2$$

$$= (r + k)^2$$

$$M_4M = r + k$$

Finally, $M_4M = NO_2 = M_4P_1 = P_7O_2 = r + k$. As we have proved, the points P_7 and N are on the circle O_2, and, the points P_1 and M are on the circle* M_4.

Q.E.D.

---

* The center M_4 of the circle M_4 is defined by the midpoint of the segment $\overline{SO_2}$. Then, the point S can be defined by a given angle TO_1O. When we draw a given angle for trisection, we must construct its vertical angle, then we draw an arbitrary circle with center O_1 to meet one side $\overrightarrow{OO_1}$ of the given angle TO_1O at the point N. At the same time, the opposite side of O_1T to meet the circle O_1 at the point S. Actually, the point S is a starting point for trisecting an angle.

III-3   Trisecting an Angle

We have discussed eleven theorems in the previous section
of this chapter. We have also investigated and proved each
theorem which had linked with an arbitrary acute angle and
its tripled angle. We have carefully and precisely examined
these basic properties among the figures including all the
related angles and sides. Also,,in the core knowledge of the
trisection, we have applied the Theorem XI to use the basic
properties of a midpoint, ratio, and geometric mean of two
segments. Consequently, the previous critical assumption in
the Chapter I has helped us to define exactly the key point
M which is the midpoint of the segment $\overline{OO_1}$, the point O is
the vertex of a given angle and the point O_1 is the vertex
of the finding angle. Obviously, the points P and O_1 can be
situated by the point M in the Theorem XI.

For a construction, we can only use an unmarked straight-edge
and compass. The degree of preciseness and accuracy, human eyes
and hands canot contest with advanced Autocad program on the
computer drawing today. Clearly, a computer is as a tool for
assisting human physical power. If we discover grandly a new
theory, a computer might be used to test or examine the validity.
But, however, a computer canot substitute human mind completely.
A new theory is valid or not, it must be based on its knowledge
system axioms, postulates, and definitions in terms of the basic
logical reasonings. On other hand, if the theory is valid, we
could always use our hands and eyes to draw a perfect figure
with two basic drawing tools and using very sharp 2H or 3H or 4H
pencils for locating a point or a line or an arc. A trisection
is not simple constraction, if a point mislocated its exact
position, then, the given angle would not be able to trisect
correctly.

Summarily, there are five steps for completing trisection in
terms of the Theorem XI. Each step must be followed the basic
constructing skills precisely, such as; a midpoint or a segment
or a line or a circle or an arc or a perpendicular line or a
parallel line.

Step 1. According to Theorem XI, we take an arbitrary
acute angle XOY (0<m∠XOY<π/2) on a geometric plane and
use two basic drawing tools to construct the given acute
angle XOY including its vertical angle as the following
Fig. 111-23.

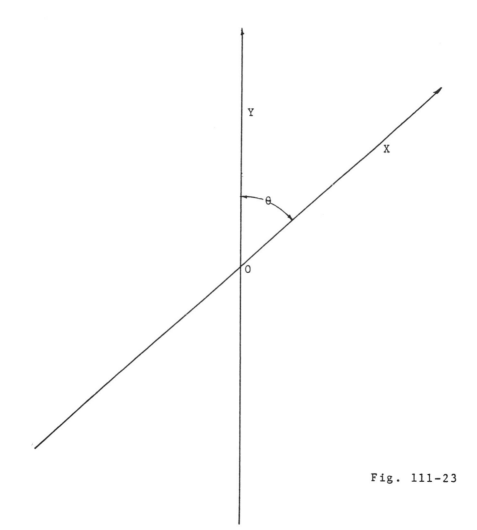

Fig. 111-23

Step 2.   Let the vertex of the given angle XOY be a center
arbitrary length be a radius, and draw two equal circles.
When we draw the first circle O to meet the opposite side of
$\overrightarrow{OX}$ at the point S and $\overrightarrow{OY}$ at the point N. Then, let the point
S to be a new center with the same radius of the circle O.
The second circle S must be passed through the center of the
first circle O and meet two different points A and B with the
circle O in the following Fig. 111-24.

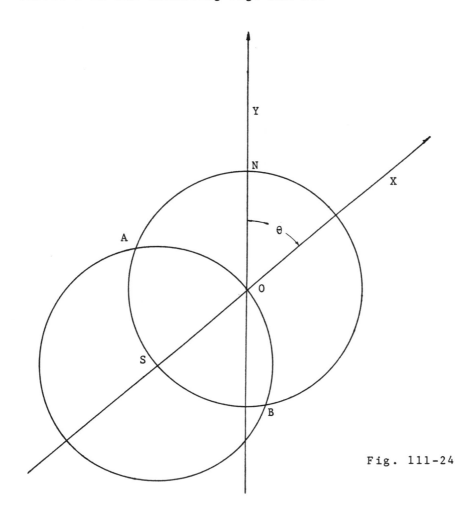

Fig. 111-24

Step 3. The point S is the center of the second circle.
We draw $\overrightarrow{SO_2}$ to perpendicular the line $\overleftrightarrow{OY}$ at the point O_2.
Continuously, we bisect the segment $\overline{SO_2}$ at the point M_1.
After that, we connect two points A and B to intercept the
segment $\overline{SO}$ at the point C, then draw the line $\overleftrightarrow{CM_1}$. The line
$\overleftrightarrow{CM_1}$ must be paralleled to the line $\overrightarrow{OY}$ as the following
Fig. 111-25.

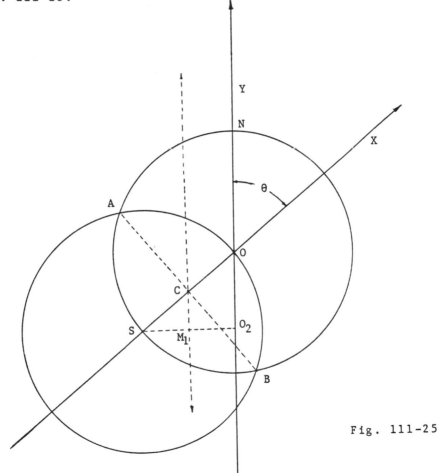

Fig. 111-25

Step 4. The fourth step is to define the key point M.
We take the length $\overline{NO_2}$ to be a radius and the point M_1 to be
a center, then we draw an arc to meet ray $\overrightarrow{OY}$ and name the
point M. After that, we let the point M be a center, the
length $\overline{MO}$ be a radius, and draw an arc to meet ray $\overrightarrow{OY}$ at O_1.
Finally, we connect $\overleftrightarrow{SO_1}$ to form the angle SO_1O which is the
one-third of the given acute angle XOY. At the same time, the
line $\overleftrightarrow{SO_1}$ will be intercepted with the circle O at the point P
and with the circle S at the point S_1. The construction is as
the following Fig. 111-26.

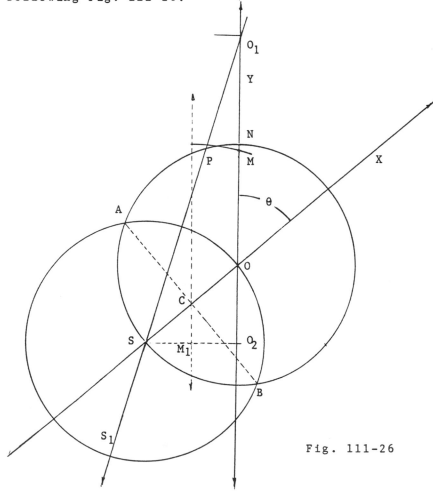

Fig. 111-26

Step 5. The final step is to complete trisection for
examing and verifying the accuracy of construction. We
extend the line $\overleftrightarrow{O_1S}$ to meet with the second circle S at
a point S_1. After that, we connect two points S_1 and O
and also extend ray $\overrightarrow{S_1O}$ to intercept the inscribed arc
of the given trisecting angle XOY. The line $\overleftrightarrow{S_1O}$ is one
of the trisecting line. On other hand, we connect two
points P and M, the line $\overleftrightarrow{PM} = \ell$ must be to meet a point
on the inscribed arc of the circle O. The point   is the
second trisecting point on the inscribed arc of the given
angle XOY. The completed construction is showing on the
Fig. 111-27. $\overrightarrow{OP_1}$ is a bisector
and perpendicular to $\overleftrightarrow{SN}$.

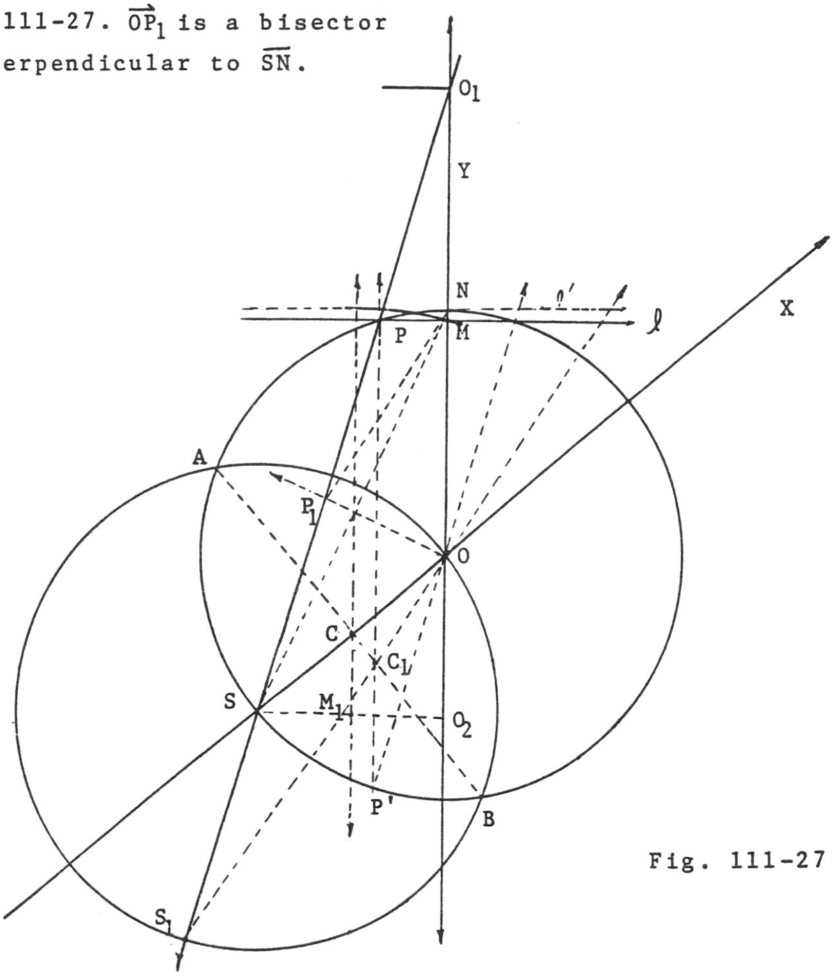

Fig. 111-27

The Fig. 111-27, an arbitrary acute angle XOY has been trisected by the five steps in terms of the Theorem XI. After the construction, it is important for us to verify the accuracy and exactitude of the trisection. The drawing can be tested or examined by each property which we have discussed in this chapter. If the figure is accurate, then we verify three key points on the drawing. First, the point C is collinear on the three lines $\overleftrightarrow{SO}$, $\overleftrightarrow{AB}$, and $\overleftrightarrow{CM_1}$, because C is the midpoint of $\overline{SO}$ and M_1 is the midpoint of segment $\overline{SO_2}$. Second, the point C_1 is intercepted by three lines $\overleftrightarrow{AB}$, $\overleftrightarrow{SO}$, and $\overleftrightarrow{PP'}$. The line $\overleftrightarrow{PP'}$ is parallel to the ray $\overrightarrow{OY}$, the line $\overrightarrow{S_1O}$ is a trisecting line to the given angle XOY. Third, the triangle OO_1S_1 must be an isosceles triangle, therefore, two segments $\overline{OO_1}$ and $\overline{OS_1}$ are congruent to each other. After that, in addition, since two circles $\odot O$ and $\odot S$ must be equal circles, hence, all radii have equal length. Finally, be sure the inscribed arc of the given angle XOY has been trisected exactly into three equal parts.

Before I took the drawing into Autocad program on a high capability in graphic computer for testing precisness, I had drawn more than 146 figures, each trisection were followed by the five steps and verified by the three testing points. All constructions canot be printed on the following pages. But, for illustration and generalization, we can take two acute angles and one obtuse angle including the $\pi/3$ angle. The trisection of $\pi/3$ angle had been shown on the the page 12 of Introduction by using computer drawing. Here, let us use two basic drawing tools straight-edge and compass. The main reason is that the $\pi/3$ angle canot be trisected by the basic drawing tools in the past mathematics history. Clearly, the earlier argument was not valid because the impossible proof did not base on the original nature of Euclidean Geometry in a mathematical model.

The following given angle XOY is defined by $\pi/3$ radian the $\pi/3$ angle has been tested by the Autocad program in a computer drawing on the page 12. Here, the $\pi/3$ angle is trisected by unmarked straight-edge and a compass under the construction of the New Theory of Trisection. As the result, the $\pi/3$ angle is divided perfectly three equal 20 degree angles.

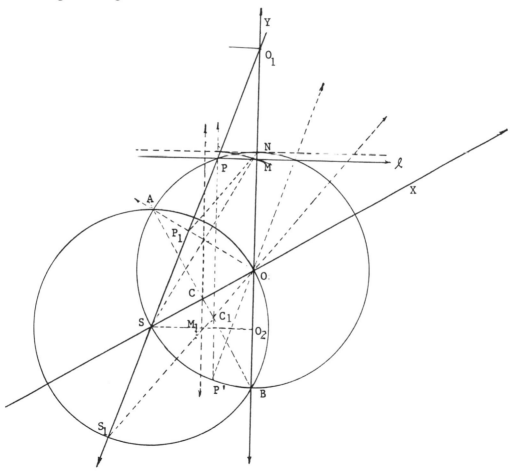

The measrement of the given acute angle XOY is between
π/3 and π/2. The angle XOY is trisected perfectly and
precisely by using two basic drawing tools.

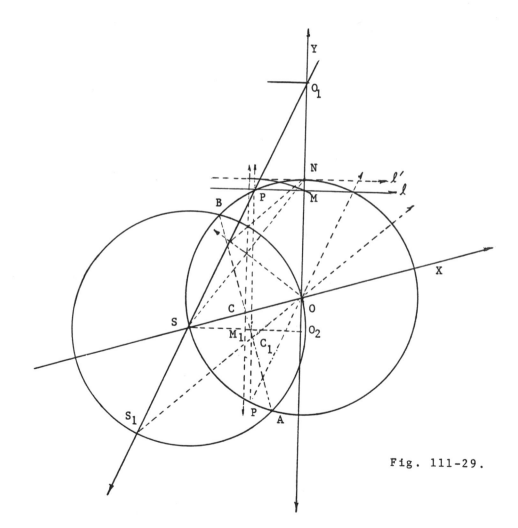

Fig. 111-29.

The measurement of the given obtuse angle XOY is between
$\pi/2$ and $2\pi/3$. The angle is trisected perfectly by using
an unmarked straight-edge and compass.

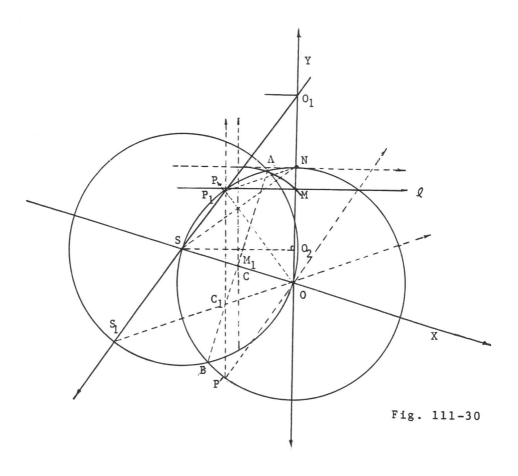

Fig. 111-30

NEW THEORY OF TRISECTION

The physical universe we live in is not only a beautiful and
intricate geometric construction, also challenges us by its on
infinite reach the philosophical, scienfitic and mathematical
reasons for us to continue to discover, explore, and develope.
In the past, we just discovered only a tiny part of the entire
universe, a lot of natural phenomena in the deeper space, we
still do not know about the true knowledge. For a difficult
problem, we canot simply say: " It is impossible to be solved."
In ancient times, Archimedes had difficulty locating a point
on an arc of a defined circle, but he did not say: "trisection
is impossible." For the entire theory, I do not dispute the
kind justice and criticisms for my work, but for the reasons
which I had stated in the preface and introduction. Evidently,
I have found reasons for the possible solution of trisection
with eleven theorems.

For the time being, I have devoted my time and mind to the
entire theory. There are three figures which were tested by
the Autocad program in the advanced computer drawing. The rest
of all the figures were drawn by my hands with two fundamental
drawing tools. Here, I canot agree more with a famous Enlish
mathematician G.H. Hardy (1877-1947) who said:" The function
of a mathemaician is to do something, to prove new theorems,
to add to mathematics and not to talk about what he or other
mathematicians have done." in his A Mathematician's Apology,
his words have encouraged me to keep going. I belive that I
did not waste my time on this study, the mathematical truth
will be proved by its logical reasonings.

We are going to end this chapter, but a new chapter is just
about begain because knowledge is endless. We have located
the vertex of one-third angle of a given angle between one-
half and one-fourth as the figures on the pages 37 and 38 show.
But we still need to situate exactly the vertex point for one-
fifth or one-seventh or one-eleventh or one-thirteenth and so
on. I will keep refreshing mind and energy on this in the
rest of my life. Also, I strong encourage all colleagues to
do and add new knowledge of mathematics

BIBIOGRAPHY

Beckman, Peter,: A History of π (PI)
  ST. Martain's Press, New York, 1971.

Bell, E.T.: The Development of Mathematics,
  Dover Publication, Inc. New York, 1940.

Boyer, Carl B.: A History of Mathematics,
  John Wiley & Sons, Inc. New York, 1968.

Brumfiel, Thomas L., Robert E. Echolz, Merrill E. Shanks, :
  Geometry, Addision-Wesley Publishing Co. Inc.London, 1960.

Bunt, Luncas N.,Philip S. Jones, Jack Bedient,: The Historical
  Roots of Elementary Mathematics, Dover Publishing, Inc.
  New York, 1976.

Cantor, George,: Transfinite Numbers,
  Dover Publishliishing, Inc. New York 1955.

Chen, Fen,: Trisector,
  United States Patent Office, Patent Number: 5,210,951.1993.

Courant, Richard, Herbert Robbins,: What is Mathematics?
  Oxford University Press, New York, 1941.

Dorrie, Heinrich,: 100 Great Problems of Elementary Mathematics.
  Dover Publications, Inc. New York, 1965.

Dulley, Underwood,: A Budget of Trisections,
  Spring-Verlag, New York, 1987.

EUCLID,: The Thirteen Books of THE ELEMENTS, Translated with
  introduction and commentary by Sir Thomas L. Heath,
  Dover Publications Inc., New York, 1956.

Hilbert, O.: The Foundation of Geometry, Translated by
  E.T. Townsend, 3rd edition, La Salle, Ill., Open Court, 1938.

Jacobs, Konrad,: Invitation to Mathematics,
  Princeton University Press, New Jersey, 1992.

Klein, F.: Famous Problems of Elementary Geometry, Translated
  by W.W. Beman and D.E. Smith, 2nd edition, New york, 1930.

Smith, David Eugene,: A Source Book Mathematics,
  Dover Publications Inc., New York, 1959.

Stewart, I.: Galois Theory,
  Capman & Hall, 2nd edition, London, 1989.

Yates, Robert C.: THE TRISECTION PROBLEM, The National
  Council of Teachers of Mathematics, U.S.A. 1971.

APPENDIX

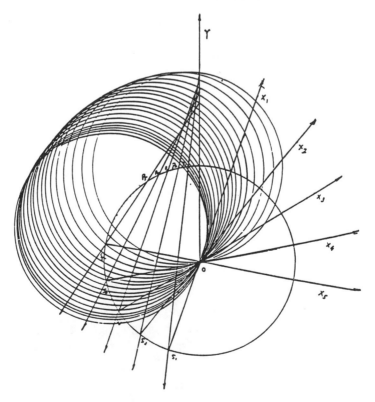

APPENDIX I. A Geometric Tree

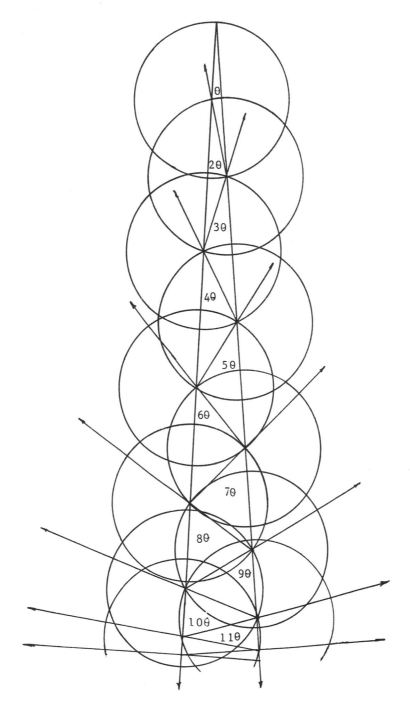

APPENDIX II. A patent Trisector

United States Patent [19]

Chen

[11] **Patent Number:** **5,210,951**

[45] **Date of Patent:** **May 18, 1993**

[54] **TRISECTOR**

[76] Inventor: **Fen Chen**, P.O. Box 16707, Alexandria, Va. 22302

[21] Appl. No.: **934,279**

[22] Filed: **Aug. 25, 1992**

[51] **Int. Cl.^{5}** **B43L 9/08**
[52] **U.S. Cl.** .... **33/1 AP**
[58] **Field of Search** .... **33/1 AP**

[56] **References Cited**

U.S. PATENT DOCUMENTS

2,222,853 11/1940 Neurohr .... 33/1 AP
3,906,638   9/1973 Romano .... 33/1 AP

FOREIGN PATENT DOCUMENTS

56135 7/1890 Austria .... 33/1 AP
41541 7/1909 Austria .... 33/1 AP

492112 6/1919 France .... 33/1 AP

OTHER PUBLICATIONS

Yates, "The Trisection Problem", 1942, pp. 7–9, 29–30, 42–44.
Washington Post, Section II. p. 1. Jan. 6, 1948, "D.C. Man Invents Device to Trisect Angle Easily".

Primary Examiner—William A. Cuchlinski, Jr.
Assistant Examiner—Alvin Wirthlin

[57] **ABSTRACT**

An instrument for trisecting an angle has two circular plates and four pointers. Two pointers are to define a given angle which can be an acute angle or obtuse angle. Two other pointers are to divide the given angle into three equal angles when they are perpendicular each other.

3 Claims, 5 Drawing Sheets

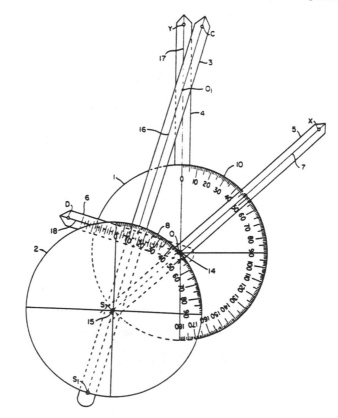

APPENDIX III.   A Pattern of Tripling Angles --- A key Point P
Moving on a Line ℓ .

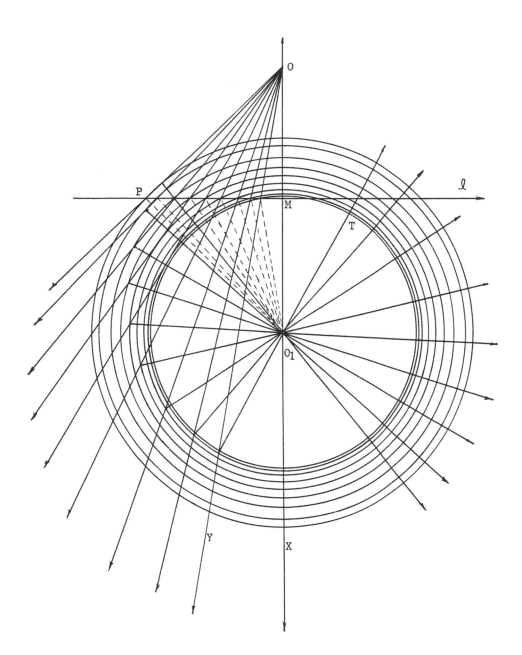

APPENDIX IV. A Pattern of Tripling Angles --- A Key Point p
Moving on a circle O_1.

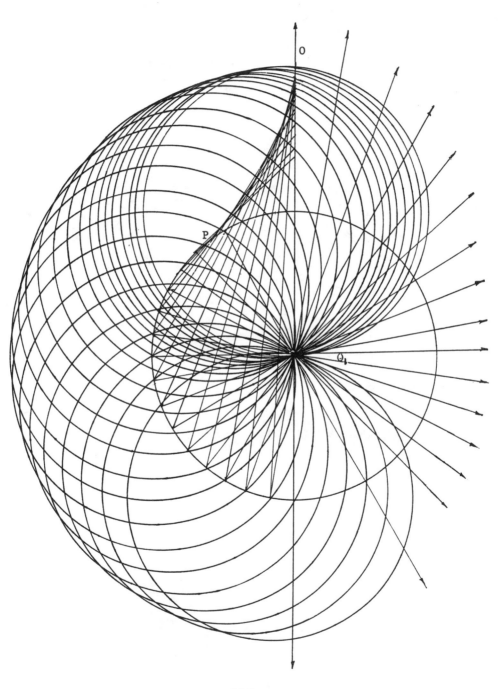

APPENDIX V. Two set circles with two fixed points N and M.

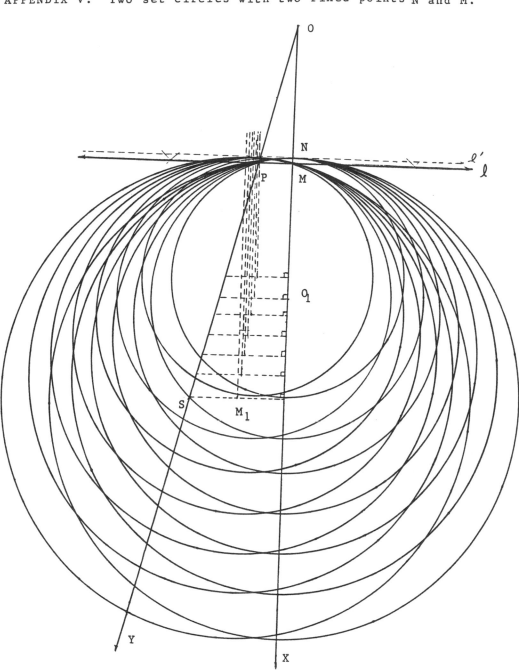

APPENDIX VI.   More Trisections --- (1)

To trisect $\dfrac{40\pi}{180}$ angle

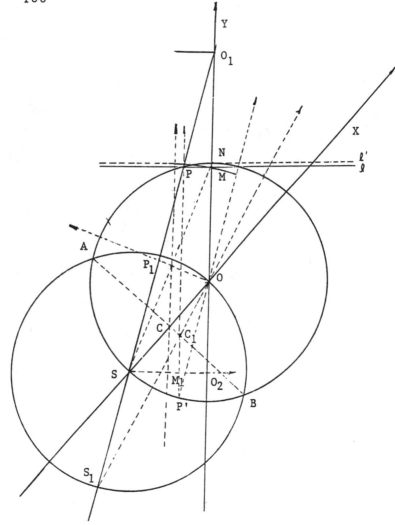

APPENDIX VI. More Trisections --- (2)

To trisect $\frac{70\pi}{180}$ angle

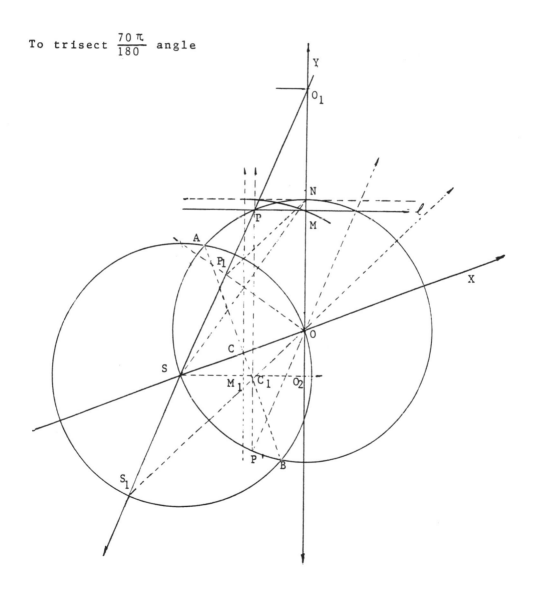

APPENDIX VI.   More Trisections --- (3)

To trisect $\frac{100\pi}{130}$ angle

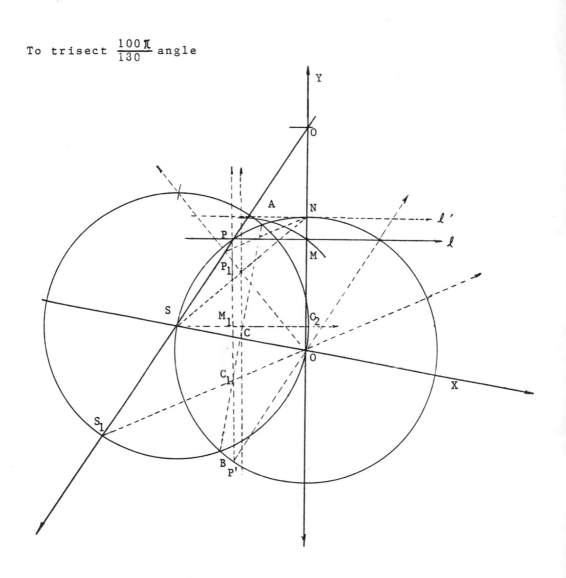

APPENDIX VI. More Trisections --- (4)

To trisect $\frac{110\pi}{180}$ angle

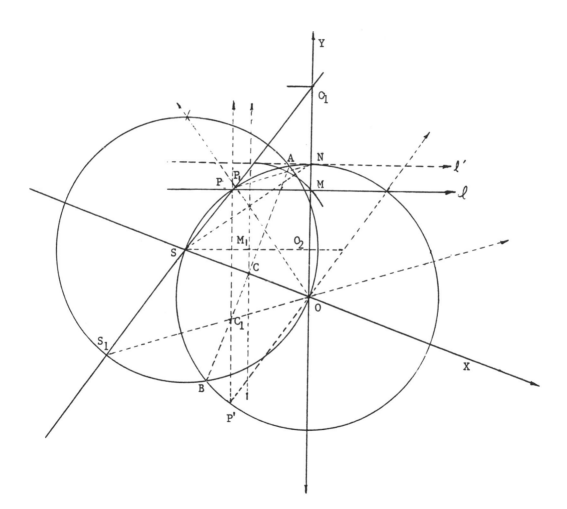

APPENDIX VI. More Trisections --- (5)

To trisect $\frac{120\pi}{180}$ angle

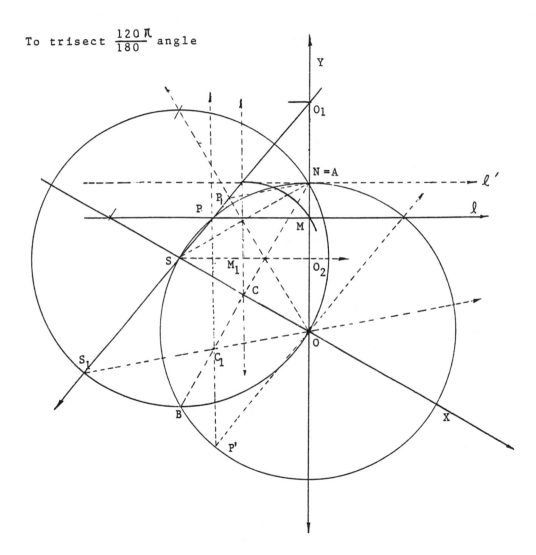

INDEX

INDEX